UFOS
and
ALIENS

UFOS

and

ALIENS

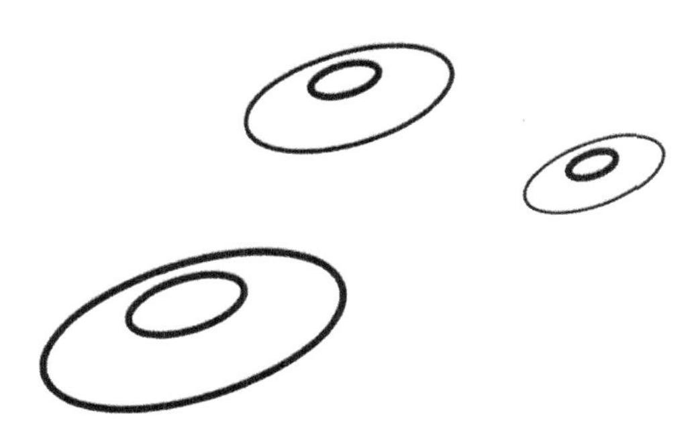

The Complete Guidebook

4th Edition

Lochlainn Seabrook

WINNER OF THE JEFFERSON DAVIS HISTORICAL GOLD MEDAL

SEA RAVEN PRESS, NASHVILLE, TENNESSEE, USA

UFOS AND ALIENS: THE COMPLETE GUIDEBOOK

Published by
Sea Raven Press, Cassidy Ravensdale, President
PO Box 1484, Spring Hill, Tennessee 37174-1484 USA
SeaRavenPress.com • searavenpress@gmail.com

1st Sea Raven Press edition: October 2005; 2nd Sea Raven Press edition: September 2010
3rd Sea Raven Press edition: December 2012; 4th Sea Raven Press edition: June 2015

ISBN: 978-0-9827700-5-4
Library of Congress Control Number: 2010933642

UFOs and Aliens: The Complete Guidebook, by Lochlainn Seabrook. Foreword by Nick Pope. Includes an index.

Front and back cover design, interior book design and layout, by Lochlainn Seabrook
Typography: Sea Raven Press Book Design
Front cover image: "Mysterious Intruder," Lochlainn Seabrook; copyright © L. Seabrook
Image on title page: "UFO Squadron," © Lochlainn Seabrook

The paper used in this book is acid-free and lignin-free. It has been certified by the Sustainable Forestry Initiative and the Forest Stewardship Council and meets all ANSI standards for archival quality paper.

PRINTED AND MANUFACTURED IN OCCUPIED TENNESSEE, FORMER CONFEDERATE STATES OF AMERICA

Dedication

To all those who devote their lives to the scientific study of UFOs.

Epigraph

The [Apollo 11] encounter was common knowledge at NASA. All Apollo and Gemini flights were followed, both at a distance and sometimes also quite closely, by space vehicles of extraterrestrial origin—flying saucers, or UFOs . . . Every time it occurred, the astronauts informed Mission Control, who then ordered absolute silence.

. . . I think that Walter Schirra aboard Mercury 8 was the first of the astronauts to use the code name "Santa Claus" to indicate the presence of flying saucers next to space capsules. However, his announcements were barely noticed by the general public.

It was a little different when James Lovell on board the Apollo 8 command module came out from behind the moon and said for everybody to hear: "Please be informed that there is a Santa Claus." Even though this happened on Christmas Day 1968, many people sensed a hidden meaning in those words.

Maurice Chatelain, former chief of NASA Communications Systems

CONTENTS

Foreword, by Nick Pope - 9
Notes to the Reader - 13
Acknowledgments - 15
Introduction, by Lochlainn Seabrook - 17

1 "High Strangeness": The Weird World of UFOs - 21
2 Aliens: Mysterious Intruders from Space - 45
3 The Past, Present, & Future of UFOs - 61
4 Taken: Alien Abduction - 80
5 Cattle Mutes & Mystery Choppers - 97
6 Lies, Denials, & Secrets: The Worldwide UFO Coverup - 109
7 Visitors in Our Fields: Crop Circles - 129
8 Black Suits, Black Cars: Meet the MIB - 133
9 UFOs Are Real: The Evidence - 139
10 Skeptics, Debunkers, & False Science - 163
11 Becoming a UFO Watcher - 181
12 Becoming a Ufologist - 193

Appendix A: UFO Sighting Report Form - 221
Appendix B: Sample UFO Sighting Report Form - 223
Glossary: 225
Index - 255
Websites (for the People Connected with this Book) - 342
Meet the Author - 343

WE

ARE

NOT

ALONE

WE NEVER HAVE BEEN

FOREWORD

by Nick Pope

The study of UFOs—or ufology, as it's widely known—can be a daunting prospect. The problem is that the subject is so vast and covers such a wide range of different, but connected, mysteries. For example, most people have heard of Roswell, where something crashed in 1947. Believers say it was an alien spacecraft, skeptics maintain that it was just a weather balloon. Entire books have been written on this single incident alone, yet Roswell is just one of a number of alleged UFO crashes, and one of literally millions of individual UFO sightings. Similarly, most people are familiar—perhaps through the Hollywood movies starring Will Smith and Tommy Lee Jones—with the concept of Men in Black. But where did Hollywood get the idea? They got it from ufology, where there have been numerous reports of sinister black-clad figures visiting UFO witnesses and intimidating them into silence. Throw in alien abductions, crop circles, cattle mutilations, and you begin to see the scope of the mystery.

How then can anyone begin to get to grips with the fascinating new science of ufology? When an internet search on words like "UFO" throw up so many millions of websites, how can people quickly get to the heart of the issue and develop a good all round understanding of the mystery? If you're reading these words, you're holding the answers to these questions in your hands, because the book you are about to read will be your guide. Lochlainn Seabrook's *UFOs and Aliens: The Complete Guidebook* is an excellent introduction to the subject and gives a comprehensive overview of the phenomenon. You will learn about the history of ufology, about all of the major cases and about some of the wide range of theories that have been put forward by believers and skeptics alike. You will learn what governments think about UFOs and how UFOs have been seen by pilots, astronauts and even presidents! You will learn about unmarked black helicopters, Area 51 and about the super-secret Majestic-12 organisation that some believe to be behind a sinister official conspiracy to keep the truth about UFOs from the public. The author even gives readers eight UFO-related questions to ask the government!

This book may set you on a journey of discovery. Many people who read books on UFOs want to get involved in the subject themselves, perhaps by looking in more depth at a particular aspect of the phenomenon. Others want to find out about the best books or websites to check out, while some want to join a UFO group and become ufologists themselves. Most books don't cover this sort of information, but this one does. In it, you'll find everything you need to know to help you follow up anything you read that interests you. You'll even find a form on which you can record any sightings you investigate, along with a checklist of how to carry out your investigations.

UFOs and Aliens: The Complete Guidebook is an informative, interesting and well-written book. Each chapter looks at a particular aspect of the UFO phenomenon, tells you how to get further information and finishes with a review of what's been covered, picking out the key points. The text is broken up into manageable, easy to read sections, and the comprehensive index, glossary, 61 real life casebook studies, and suggested reading section mean that you'll be able to navigate your way around this book easily and find out all you need to know about how to take your interest further, whether through other books or on the internet. Although primarily aimed at people new to the subject, there are many seasoned ufologists who would do well to read this book and learn from it.

I used to run the British Government's UFO Project, working at the Ministry of Defense. You'll find out a little bit about me in this book. I ran the British version of the study that the United States Air Force used to call Project Blue Book. I came into the job as a skeptic, but after three years of official research and investigation into UFOs, alien abductions, crop circles and other strange phenomena, I came to believe that we should not dismiss the possibility that some UFOs really might be extraterrestrial spacecraft. Why do I make this extraordinary claim? I make it because while most UFOs can be explained as misidentifications of aircraft, meteors, satellites, weather balloons, etc, a small percentage defy any conventional explanation. Some UFOs are seen by trained observers such as police officers, military personnel and pilots. Some UFOs are capable of speeds and manoeuvres that go way beyond anything that even the most advanced aircraft and drones can do. Some UFOs are tracked on radar. Sometimes UFOs are captured on camera or video, and the pictures are analysed scientifically, and can't be debunked. Occasionally, UFOs land, and evidence is found on the ground, such as unusually high radiation levels. This is the sort of evidence that convinces me that UFOs should be studied carefully and this is exactly the sort of evidence that you will shortly be finding out for yourself, as you read this book. Now, read on, and begin your voyage of discovery into the greatest mystery of our age. Prepare to be amazed.

Nick Pope
September 2010
London, United Kingdom

Author, journalist and TV personality Nick Pope used to run the British Government's UFO project at the Ministry of Defence (MoD). Initially sceptical, his research and investigation into the UFO phenomenon and access to classified government files on the subject soon convinced him that the phenomenon raised important defence and national security issues, especially when the witnesses were military pilots, or where UFOs were tracked on radar.

While working on the MoD's UFO project Nick Pope also looked into alien abductions, crop circles, animal mutilations, remote viewing and ghosts. He is now recognised as a leading authority on UFOs, the unexplained and conspiracy theories. He does extensive media work, lectures all around the world and has acted as presenter, consultant or contributor on numerous TV and radio shows.

Notes to the Reader

Our fast-paced, information-rich, highly changeable society can make it difficult to stay up to date with the latest names, postal addresses, telephone numbers, email addresses, and Websites. Because of this I can't guarantee that all of the data in this book (particularly in Chapter 12) is current. Check the Internet for the latest information.

Key to symbols used throughout this book:

Bolded words direct the reader to the glossary.

This symbol indicates an actual casebook study.

This symbol indicates examples, or highlighted material.

This symbol indicates a review of material.

This symbol directs the reader to additional source material.

This symbol indicates additional source material.

This symbol indicates the end of a chapter.

Reality or fiction? Read this book and judge for yourself. Photo © Lochlainn Seabrook

ACKNOWLEDGMENTS

Special thanks to my wife Cassidy and my daughters Fiona and Dixie, and to Nick Pope, Stanton T. Friedman, Timothy Good, Dr. Bruce Maccabee, and Erin Ryder. Their support and generosity is much appreciated.

"Lochlainn Seabrook's *UFOs and Aliens: The Complete Guidebook*, is a truly remarkable introduction to UFOs that transcends all other guidebooks in the genre." — ERIN RYDER, from SyFy Channel's hit TV series *Destination Truth* and National Geographic's popular series *Chasing UFOs*. Photo copyright © Erin Ryder/Sea Raven Press.

INTRODUCTION

Updated for the 4th edition

Knowing that so many people are fascinated with UFOs and aliens, and naturally and spontaneously believe in and accept their reality, I felt it was time to devote a guidebook to these topics. Not one written by a skeptic, debunker, or atheist (there are already a number of these on the market), but one written by a true believer using a scientific approach, so that the subject is treated in an open and fair manner. Thus, in this book you will find no prejudice toward the idea of UFOs and aliens, no ridicule toward those who believe in them, have seen them, or experienced them in some way. Quite the opposite.

I treat the topic of this book as if it is real, because it is real! Like millions around the world, I've had my own deeply personal and life-altering experiences with UFOs and aliens, experiences that are just as authentic to me as our everyday reality here on Earth. One of these, in fact, occurred while writing the first edition of this book.

On August 1, 2005, I was taking a business flight across the U.S. It was about 11:00 in the morning and we were at 39,000 feet over Charleston, West Virginia. I was gazing casually out the window of our Boeing 737 at the line where the clouds and sky meet (just below us). As my eyes ran along this line I spotted a brilliant silver sphere sitting right above it, approximately three to five miles distant.

The object was level with our plane and stationary and seemed to be rotating. As it hovered it glimmered and sparkled, throwing off great flashes of light. It was either self-luminescent or reflecting sunlight, and was absolutely dazzling. Before I could get my camera in place to take a picture the strange object blinked off, vanishing into nothingness.

On another occasion, while I was living in south Florida, I saw something that would change anyone's worldview. On a clear quiet night on Key Biscayne, I happened to be looking across the water at the lights of Miami a few miles away, when I saw a gleaming white orb tear across the horizon about a 1,000 feet above the city. It flew perfectly level (not a meteorite), and, after covering at least 20 miles in only a second or two (not a human-made aircraft), it flashed once and disappeared. Was I hallucinating? Hardly. My wife, who was standing next to me, saw it too.

Of the many other incidents I could recount, a more recent sighting bears inclusion here.

In the Spring of 2014 we were walking in an uninhabited public area near our home outside Nashville, when we saw a small aerial object appear suddenly from out of a thick forest 100 meters away. It was about 18 inches in diameter and looked like a large, dull, grayish aspirin standing on its end. It was about 10 feet off the ground, and had what looked like a short "string" hanging from the bottom.

Assuming it was a helium-filled party balloon that had gotten loose, we expected it to simply drift up into the sky and float away.

Instead, it traveled against the wind, moving very intentionally along the tree line, maneuvering adroitly through a complex tangle of branches without touching any of them. This went on for several minutes. It was at this point that we realized this was not a balloon, but a craft being controlled by some form of intelligence.

As it moved slowly up and down through the trees along the edge of the woods, we got the feeling it was looking for something—when all of the sudden it seemed to become aware of us. We watched in amazement as the object immediately turned and began coming, edge on, right at us, on an exact course. Still not sure what it was, we didn't panic, but merely watched in stunned silence as it sailed just 20 feet above our heads, as if to inspect us. Still flying perfectly straight against the breeze, it continued steadily past, gradually climbing in altitude until it became a dot in the far distance. It then disappeared into the daytime sky, by now several thousand feet in the air.

As it traveled overhead, we had gotten a detailed look. The "string" on the bottom was solid (it never moved in the wind). Perhaps a disguised broadcast-receiving wire? Its "skin" looked rumpled, like a balloon that had lost some of its air. Yet the surface was solid, never once bending or flexing like a deflated balloon would. In fact, close up the covering appeared metallic. A hard-shell protective case for the electronics inside? Lastly, it acted as if it wanted us to see it, to know it was there. Not only had it made no attempt to conceal itself, it quite deliberately showed itself to us. Neither its appearance or its behavior matches any known manmade object. In fact, it felt eerily nonhuman.

After the feelings of exhilaration subsided we matter-of-factly discussed what we had seen: a bizarre aspirin-shaped object masquerading as a child's toy balloon. But what was it? A miniature probe of some kind, reconnoitering the area for biological material? We'll never know. We can't say for sure that it was alien. But why would humans make such an object when it would be much easier, and less costly, to simply travel to the area themselves?

These are only three of the many unidentified flying objects that I've encountered. After such experiences one can never be completely skeptical again.

If you doubt the reality of extraterrestrials and their otherworldly craft, this book will be a revelation. For *UFOs and Aliens: The Complete Guidebook* is an exhaustive compendium of nearly all of the currently available information on this fascinating subject. It may make a believer out of you too.

Lochlainn Seabrook
June 2015
Franklin, Tennessee, USA

UFOS

and

ALIENS

The Complete Guidebook

1

"HIGH STRANGENESS": THE WEIRD WORLD OF UFOS

"I can assure you that flying saucers . . . are not constructed by any power on earth." — President Harry Truman, Washington, D.C., 1950

THE UBIQUITOUS UFO!

Everyday, every hour, somewhere in the world, someone sees a UFO. From the U.S. to the Pacific Islands, from Norway to South Africa, from Canada to Argentina, from Russia to Australia, UFOs have been sighted—and continue to be sighted—thousands of times a year in every country on Earth. Usually not by astronomers, but by ordinary people like you and me.

So many are seen that it's been estimated that for every UFO sighting that's reported, ten go unreported. Clearly, something is going on, but what?

> UFO FACTOID
> UFOs are everywhere. There could be one hovering or flying over your home right now.

UNDERSTANDING UFOS

Let's begin with the most obvious question: just what is a UFO? Since there's so much confusion about them, let's define what they are.

First, what does UFO mean? UFO is an acronym for **Unidentified Flying Object**. The word UFO is really just a handy way of referring to something that's up in the sky that we can't identify.

NOT EVERY UNKNOWN THING IN THE SKY IS A UFO

This doesn't necessarily mean that every unknown object that's flying around in the heavens is a UFO. Actually, most of the time what seems to be a UFO turns out later to be an IFO: an Identified Flying Object.

Many everyday objects have been mistaken for UFOs at one time or another. Some of these are artificial (created by humans); others are natural (created by Nature).

Among the things that have been mistaken for UFOs are clouds, weather balloons, party balloons, birds, searchlights, lighthouses, meteors, gases, lightning, Saint Elmo's Fire, emergency flares, blimps, airplanes, satellites, passenger jets, helicopters, Chinese lanterns, discarded space junk, secret experimental military aircraft, refracted light, hot gases, temperature inversions, auroras, ion clouds, barium clouds, laser displays, radio waves, stars, and even bright planets, like Venus, Mars, and Saturn.

UFO FACTOID
Sometimes a cloud can look like a UFO. Other times UFOs hide themselves in clouds or even disguise themselves to look like clouds.

Sometimes **mirages** and **optical illusions** make people think they're seeing a UFO when they're really not. The Sun can look like a UFO under the right conditions, as can the Moon. Kites, car headlights, streetlights, and even ordinary houselights, have also been misidentified as UFOs.

Rotting vegetation can create methane gas that glows in the dark (particularly on moist nights), giving off wavy lines of colored light. This type of vapor, called "swamp gas," or Will-O'-the-Wisp, has been the source of some UFO sightings.

One of the most common objects mistaken for UFOs is the Chinese lantern (paper balloons that rise from the heat created by candles inside them). Hoaxers sometimes purposefully prank the public by sending these colorful paper lights up into the night sky, especially where UFOs are often seen. At heights over 1,000 feet it's difficult, if not impossible, to tell what the queerly glowing, silent objects are, thus they're quickly dubbed "UFOs" by unsuspecting eyewitnesses.

Some companies are now actually making large balloons that are shaped exactly like UFOs (usually disk-shaped), and which come with a bright silvery coating that glints in the Sun, giving them a solid metallic appearance. As the helium-filled balloons soar and roll through the sky, only the most highly trained observers would ever be able to tell that they're not objects from another world.

Adding to the confusion, the human body itself can create things that people mistake for UFOs. One of the more common of these is the "vitreous floater." This is a dead cell in the fluid of the eyeball that can dart around in one's field of vision, making it look exactly like a UFO moving around in the sky. Dust and dirt on one's cornea can produce the same effect. Reflections from windows and eyeglasses too can sometimes create UFO-like images.

Another example is what's called "autokinesis." This occurs when you stare at a stationary object, like a star, for a long time without anything to compare it to (such as a telephone pole, a building, or a hill). Eventually the star will seem to move. But in this case it's not the star that's moving, it's your eyes. Scientific studies show that the human eye is constantly and involuntarily moving about, even when a person is sound asleep.

> UFO FACTOID
> While many people mistake ordinary objects for UFOs, keep in mind that at least 10 percent of *all* sightings are genuine unidentified flying objects.

But none of these things are true UFOs, because we know what they are. By carefully investigating these specific types of phenomena we can always identify them. Being skeptical but open-minded is good science.

THERE ARE GENUINE UFOS!

At least 10 percent of the time (or one out of ten times), however, we can't tell what the object in the sky is—no matter how hard we try, no matter how long we take, and no matter how much we study it. This is a genuine true-to-life UFO.

This may sound like a small number. But keep in mind that 10 percent of all sightings amounts to tens of thousands of actual unidentified flying objects; mysterious craft flying around in our airspace that even our own governments—with all of their highly trained military personnel and highly educated scientists—can't explain.

WHO SEES UFOS?

How many people have seen a UFO? As of 2007, 14 percent of the American population have reported seeing a UFO. That's 42 million people in the U.S. alone.

Just what kind of a person sees UFOs? Though some studies show that you're more likely to see a UFO if you're intelligent and educated, UFO witnesses include a complete cross-section of humanity. People who see them come from every race, income level, religion, and country.

Even a person's age is no barrier: many UFOs are observed by young people. UFOs are so common today that if you haven't seen one, you probably already know someone who has. Ask around among your friends, family, and coworkers. You might be surprised.

WHERE ARE UFOS SEEN?

UFOs don't seem to prefer one place over another. So if you want to see one, it doesn't matter whether you live in a hot tropical area or a cold wintry one. UFOs can appear anywhere and everywhere, and as we'll see, they have.

> UFO FACTOID
> People have been seeing UFOs for a long time. A European tapestry from the 12th Century shows a hat-shaped object flying through the sky in the background.

UFOs have been spotted by people sitting on their porches, by passengers on ships and planes, by people driving down country roads and busy expressways, by people living near the sea, in the desert, in the mountains, and in snow-covered forests.

They've been seen by people in the middle of the world's most heavily populated cities, and in the middle of the world's most uninhabited regions. In short, wherever there are people UFOs have been sighted.

WHEN ARE UFOS SEEN & HOW ARE THEY CLASSIFIED?

Amazingly, UFOs don't seem to be affected by Earth's weather. So they can be seen at any time of the year and in all seasons.

And even though you might think otherwise, not all UFOs are seen at night. Thousands of UFOs are seen every year during the day. These are called "daytime disks," or "diurnal disks"—**DDs** for short.

But most UFOs are still seen at night, probably because they're often easier to spot in the dark. UFOs that appear after the Sun goes down are called "nighttime lights" or "nocturnal lights"—**NLs** for short.

Interestingly, some UFOs are not actually seen by the human eye at all, no matter what time of day or night it is. Instead these are seen by an electronic eye called radar, a common instrument used by airports and military bases. This type of UFO, called a "radar-visual," or **RV** for short, shows up as an unidentifiable "blip" on a radar screen.

In summary the three basic visual categories of UFOs are: DDs, NLs, and RVs.

UFO FACTOID
While most UFOs show no sign of a power source (e.g., an engine) or of any kind of propulsion (e.g., propellers), some possess large spheres on the underside, believed to be a type of electro-magnetic power unit.

WHAT SIZE IS A UFO?

UFO reports show that unidentified flying objects come in every conceivable size, with the smaller ones probably being used for reconnaissance and the larger ones for long distance, or interstellar, travel.

The range in actual size boggles the mind. Some UFOs are as small as a dot of light (at close range). Some are the size of a ring or a marble. Some are the size of a golf ball, tennis ball, or a basketball. Some are the size of a bathtub, a telephone booth, or a refrigerator. Others have been seen that are the size of a car, truck, or small airplane, while still others appear as big as a garage, a house, or a large store.

A number of people have witnessed UFOs that were the size of a jetliner, or even an aircraft carrier. And some have seen "super UFOs," or "mother ships," giant unidentified craft that are as big as five football fields linked end to end.

But some UFOs are even larger than this. Here are several examples from the files of authentic UFO reports.

Casebook Study 1: On June 11, 1985, the pilots and passengers aboard a Boeing 747 flying out of Peking, China, all watched in awe as a UFO streaked past in front of them. The pilots, professionally trained to judge distances, stated that the bright oval object was six miles wide.

Amazingly, there are UFOs of even greater size.

Casebook Study 2: In January 1985 Russian officials reported the sighting of an unidentified object by commercial pilots. It first appeared as a yellow star, but as it drew closer to their airliner it began to change shape, turning into a large "green cloud," inside which the pilots could see flashing colored lights.

Now flying alongside the airliner the bizarre craft continued to shape-shift into a wide variety of forms while at the same time sending out blinding cone-shaped beams of light in all directions.

By now frantic passengers could also see the weird threatening object. The pilots tried to calm them, but it didn't work. For as the airliner flew over several lakes, the size of the UFO could now be easily determined by comparing it to the known size of the bodies of water below: the UFO was about twenty-five miles long!

UFO FACTOID
While UFOs most often appear like two saucers placed together, actually they can look like almost anything.

WHAT IS A UFO SHAPED LIKE?

UFOs come in every shape, although it's not always easy to tell exactly what that shape is. Part of the problem is that UFOs often travel so fast that they're hard to see clearly.

But the main difficulty is that most UFOs seem to produce very powerful gravitational fields. These force fields bend light waves, causing the UFO to look fuzzy or blurry.

Luckily this type of distortion is rare. Most of the time witnesses get a good clear view of the shape of the object as it passes by. And what they see is truly unbelievable, as the following list—compiled from authentic UFO reports—shows.

People have seen UFOs shaped like acorns, airfoils, airplanes, ashtrays, balloons, balls, bananas, bells, birds, boomerangs, bowls, boxes, bullets, chairs, cigars, circles, clouds, coins, crescents, crosses, cubes, cucumbers, cups, cylinders, diamonds, doughnuts, dumbbells, eggs, footballs, garbage cans, half-circles, hats, hexagons, horseshoes, ice-cream cones, ironing boards, jellyfish, kites, lanterns, mattresses, mushrooms, oil drums, ovals, pans, pencils, pies, plates, polyhedrons, pyramids, rings, rockets, scythes, soap bubbles, spheres, spindles, spirals, squares, stars, stools, teardrops, tetrahedrons, tops, torpedoes, trapezoids, trumpets, tubes, walnuts, water spouts, and wings.

Of course, the best known shape for a UFO is the saucer or disk. This is considered the "classic" UFO design, one that's been observed for literally thousands of years.

And yet today, for reasons that no one fully understands, the triangle-shaped UFO is one of the most prevalent types that's seen. Perhaps this is because the beings that make UFOs are modernizing their spacecraft each year, just as humans do. Or perhaps these beings are just trying to imitate the triangular aircraft that we're now making so they can fly around without being noticed. Whatever the answer, it's a fact that triangular or wing-shaped UFOs were being reported long before we invented the airplane.

UFO FACTOID
Adding to the mystery of UFOs is the fact that many of them seem to be able to shape-shift, taking on a variety of forms, from boxes to orbs, from diamonds to cigars, from triangles to crescents, all in a moment's notice.

The idea that aliens are probably imitating our aircraft isn't as strange as it sounds, for another unusual aspect of UFOs is that some of them seem to be able to shape-shift. That is, they can change the way they look any time they want. Indeed, some people have reported seeing UFOs transform themselves into airplanes and then back into UFOs again.

Other people have spotted UFOs that can alter their dimensions, shrinking from the size of a house to a small point of light, and then quickly expanding again to their original size. Still others have seen UFOs that went from a saucer shape to a box shape to something like a cluster of balloons, all as they flew at breathtaking speeds.

HOW MANY TYPES OF UFOS ARE THERE?

UFOs can be divided into two basic categories: 1) those that fly in the air, and 2) those that travel underwater. As this book will be mainly devoted to aerial objects, let's take a brief look at category number 2.

Though not as well-known as their airborne cousins, more and more people are seeing what's called a **USO** (an "unidentified submerged object") or **UUO** (an "unidentified underwater object"). Like UFOs, USOs seem to defy the laws of physics: they've been tracked traveling at depths of 20,000 feet and at incredible velocities, some at nearly the speed of sound (roughly 700 mph).

Our typical military submarine, however, can only go to a depth of about 4,000 feet and travels at a top speed of only twenty-nine mph! Just as astonishing many USOs also possess the dual ability to move both underwater *and* through the air, something we humans have not come close to achieving—with any type of craft.

UFO FACTOID
The first USO sighting was reported thousands of years ago. Ancient peoples described them in the vernacular of the time as "plates," "spheres," "fire," "swords," "shields," and "globes."

Despite their stunning technology USOs are not a modern phenomenon. One of the first ever recorded occurred in India over 300 years before the birth of Jesus. In the year 329 BCE, Alexander the Great spotted what he described as shimmering "flying shields" that flew in and out of a river he was trying to cross with his army. The silvery objects dove at the men, "spitting fire from around their rims," terrifying everyone, including Alexander's animals.

Another early USO incident of note took place on October 11, 1492, in an area now known as the Bermuda Triangle. Christopher Columbus and Pedro Gutierrez were standing on the deck of the *Santa Maria* when unusual lights were seen flickering under the water around the ship. As the two explorers watched in amazement, a disk-shaped object rose up from out of the ocean waves emitting brilliant flashes of light.

Here, from the archives of genuine UFO reports, are two examples of modern USO sightings.

Casebook Study 3: In the summer of 1942 an Australian fighter pilot from the **RAAF** watched in shock as a luminescent bronze UFO, 150 feet long and fifty-feet wide, came up and flew alongside his plane for several minutes. The craft then turned and dove straight down into the ocean below, throwing up huge waves.

Casebook Study 4: Arguably the most famous USO case to date is the Shag Harbour UFO Incident: on October 4, 1967, over a dozen people watched as a strange, sixty-foot wide, caterpillar-like UFO with flashing lights, flew across the sky, then plunged into the waters off southern Nova Scotia, leaving an orange foam on the surface. It then submarined a few dozen miles up the coast, where it was joined by another underwater object off the town of Shelburne.

> UFO FACTOID
> USO sightings are becoming more common, by both civilian and military eyewitnesses.

The Canadian government tried to cover up the affair, but twenty-five years later, declassified **X-files** and eyewitness testimony revealed that the military had sent a small fleet of ships to the area in an attempt to identify and capture (or destroy) the weird craft. They were tracked on both **radar** and **sonar**, and divers were sent down, later confirming that the two objects were indeed USOs, and "not from our planet."

Eventually the USOs powered into the Gulf of Maine, rose to the surface of the ocean and sped straight up, disappearing into the sky. As we don't have the technology to build vehicles that can both fly in outer space *and* travel underwater, it's obvious that these USOs were extraterrestrial.

Besides military and civilian eyewitnesses, commercial pilots had also observed the event, making this one of the best documented sightings in history. Still, to this day the Shag Harbour UFO Incident, referred to as "Canada's Roswell," remains an enigma, for no one knows what the objects were or where they originated. All we know for sure is that they weren't ours.

WHAT COLOR IS A UFO?

While unidentified *flying* objects have been seen in nearly every color of the rainbow, there are a few basic colors that are most common. DDs are usually metallic gray or silver in color, while NLs are typically bright white or silver.

Other colors that UFOs have been seen in include: red, yellow, blue, bronze, purple, turquoise, maroon, pink, green, ivory, pearl, pure white, crimson, gray, orange, gold, flame-colored, and amber. UFOs have even been seen in colors that we have no words for.

Sometimes a UFO's color can be soft and hazy (like the Moon when seen during the day), and other times a UFO can be so bright and intense that it hurts to look at it. But others sport pleasant pastel shades, like soft pinks and purples.

UFO FACTOID
UFOs come in every imaginable shape, size, and color, probably for the purpose of performing a multitude of different functions.

Some UFOs look like dull metal, similar to an aluminum pan. Others flash and glint like highly polished silver, ice, or glass. Some are clear or translucent, so you can see right through them—only their outlines show.

Often UFOs are bicolored or even multicolored. For example, people have seen UFOs that are red on top and blue on the bottom; or silver on the sides, gold on top, and orange on the bottom. Many UFOs have no color at all. They're completely black, probably a cloaking device related to stealth.

Still others seem to be self-luminescent; that is, the entire craft simply glows in one or more colors. Some people have reported seeing UFOs that radiate an ever-changing kaleidoscope of colors all at once.

UFOS & THEIR LIGHTS

Identifying the color of a UFO itself is sometimes made more difficult because of the fact that most of them possess multicolored navigation lights.

For instance, some UFOs have a single bright strobe or flashing white light on the bottom; others have weird reddish domes on top; others have a mixture of colored lights all around the edges, and on the top and bottom as well. Some of these flash irregularly, others stay lit continuously. Furthermore, those lights that pulsate often do so in time with one another, or in predictable patterns, while at other times they'll flash randomly in no set sequence.

One of the more peculiar aspects of UFOs (and there are many) is that there seems to be no definitive pattern to the colors or the locations of lights on the outside of the craft. Every combination has been seen, from UFOs that are completely covered with lights, to UFOs with no lights.

Though most UFOs have at least a few running lights on them, others have what can only be called "spotlights" or "searchlights." Often these intense beams of light shoot out from the front or from the bottom of the object. People who've seen UFOs with spotlights say that they seem to be looking for something on the ground. But what? No one knows.

UFOS & THEIR WINDOWS

How does somebody traveling in a UFO see outside? If you think the answer is "through a window," you're only partially right. This is because while some UFOs do have windows, many don't. In fact, most UFOs are completely windowless. So how do the occupants see out?

> UFO FACTOID
> The average UFO has no windows. Those that do sport windows of nearly every possible configuration, from tiny portholes to large "picture windows" through which the strange occupants can sometimes be seen.

First, many UFOs, like RPVs ("remotely piloted vehicles"), have no occupants at all. These drone UFOs, or UAVs ("unmanned aerial vehicles"), are similar to our satellites; or more precisely, like our drone aircraft, such as the MQ-9 Reaper, the RQ-4 Global Hawk, and the RQ-2 Pioneer: they're pilotless machines that are sent out to gather data. No living being needs to be on board. Of course, such UFOs, called "probes," don't need windows.

Second, even when a UFO has occupants in it a window isn't necessarily needed. For many types of UFOs are outfitted with navigation systems, perhaps similar to—but far more advanced than—our radar, which allow the crew members to steer through the air without having to see outside. They simply look at the information on their electronic screens to go where they want to go.

This is easier than you might imagine. Modern submarines operate without any windows at all. And yet using radar, their crews can navigate thousands of miles underwater (staying down for months at a time) without ever looking out or coming to the surface.

Some UFOs, however, seem to need windows, because many of them do have them. Saucer-shaped UFOs are often seen with domes on the top and little portholes going all the way around the edge. Diamond-shaped UFOs often have banks of windows along their sides.

Some rectangular UFOs have been spotted with huge "picture

windows" in them. These windows are often so big that eyewitnesses can see control panels and even the living beings inside. In one case the strange-looking alien pilots waved at the humans as they flashed past.

THE SOUNDS OF UFOS

UFOs have been known to make many different sounds, from a low soft rumbling that rattles the windows in a house, to a shrill high-pitched whine that can hurt a person's ears. Some people report that they've heard UFOs make a sound like a jet or a rocket taking off. Others have heard them hiss, hum, whine, beep, whistle, and even make buzzing sounds like bees. One UFO was heard that made a curious musical tone.

Most baffling though is that many UFOs make no sound at all. Some of the biggest UFOs, the "super UFOs," are often completely noiseless as they glide slowly over homes and military installations. Even during those maneuvers when you'd expect to hear the sound of an engine—such as when a UFO reverses, accelerates, or hovers—witnesses report that there's often not a single noise of any kind coming from the craft. Just complete silence.

This is surely one of the many pieces of evidence which shows that we're dealing with an alien, or off-Earth, technology. For humans have yet to invent an aerial machine that can fly straight and level at low or high speeds, or hover, without making noise.

As a matter of fact, we can't even make a silent lawn mower, let alone a silent aircraft. If we did have this kind of technology, the U.S. would certainly be leading the rest of the world in every area, particularly militarily and economically. The truth is we aren't: according to some studies, each year America is actually lagging further behind other nations in these two areas.

> UFO FACTOID
> The majority of UFOs are silent. We humans have yet to construct a completely noiseless machine—of any kind.

THE BEHAVIOR OF UFOS

The way in which UFOs behave is one of the most mystifying aspects of these incredible craft. For instance, many UFOs have been seen traveling alone. Just as many have been seen flying in large numbers, like a flock of birds flying in tight, highly disciplined echelon formations.

But they've also been spotted gyrating through the air uncontrollably.

From such reports alone it's clear that UFOs don't act the same way all the time, and that there's an enormous variation in the manner in which they behave. This is not too surprising: since UFOs have so many different shapes and sizes, we wouldn't expect them all to fly, perform, or move the same way.

UFO FACTOID
Unlike human-made aircraft, UFOs behave in strange and unfamiliar ways, even seemingly defying the known laws of physics.

One of the more common behaviors is for a UFO to simply hover in the air, like a helicopter—only without sound. Some move in a strange hopping motion, as if they're stones skipping over water. Others tremble, wobble, vibrate, throb, and shudder. Some sway back and forth like a falling leaf. Others swing from right to left, like a pendulum on a clock.

Some dance, dart, jump, zig-zag, bob, and whirl crazily in the air, weaving circular patterns of light in the sky. Others dive and climb, accelerating rapidly with blinding speed into the clouds. Still others are like tourists who enjoy sightseeing on Earth, moving at what can only be called a leisurely pace, completely unconcerned with who observes them.

Some UFOs travel like bullets, at tremendous velocities—so fast that it's almost impossible to see them. A number of them have been clocked on radar moving more than 12,000 miles per hour (mph) across the sky. A UFO was measured at the White Sands Proving Grounds in New Mexico traveling at 18,000 mph. Still others have been tracked moving at between 60,000 and 100,000 mph.

UFO FACTOID
The Space Shuttle can achieve the same high speeds as some UFOs. But only in outer space.

But apparently even this isn't the fastest speed at which UFOs are able to travel. Air Force intelligence officer Steve Lewis studied UFOs for the military for twelve years and concluded that they use some kind of an advanced propulsion system, one that allows them to literally travel at the speed of light. That's 186,000 miles per second, or 669,500,000 mph!

Of course in Earth's atmosphere even the slower speeds at which UFOs travel are much faster than the current fastest human-made aircraft: the X-43A Scramjet, a one-of-a-kind, unrecoverable, unmanned vehicle that got up to a mind-boggling 6,600 mph on November 16, 2004 (after which it was purposefully crashed into the Pacific Ocean). Even then the X-43A was only able to maintain this speed for eleven seconds. UFOs travel at much higher speeds than this for much longer periods.

Actually, most human-made aircraft go far slower than the X-43A Scramjet. A standard military jet flies at around 1,200 mph, a typical commercial jet liner travels at about 600 mph, and a small private airplane only goes about 150 mph.

It's true that the American space shuttle can reach speeds of 17,000 mph, and our fastest spacecraft, Voyagers 1 and 2, the robotic space explorers launched in 1977, traveled along at a stunning 38,000 mph. But the shuttle and Voyager probes can only travel at these speeds in outer space where there's no air, and thus no friction to slow them down. UFOs travel at many times these speeds right here in Earth's atmosphere, sometimes very close to the ground. How they do this is a mystery we haven't begun to unravel. Our earthbound aircraft would explode, burn up, and disintegrate long before reaching these types of velocities.

UFO FACTOID
The Voyager Probes 1 and 2 are humanity's fastest objects, traveling at nearly 40,000 mph in outer space. Yet UFOs can move at much greater speeds, even in Earth's atmosphere.

Another astonishing behavior of UFOs is that they can make ninety-degree turns at high speeds. There is no human-made aircraft that can turn a corner like this while flying through the air, particularly at speeds of up to hundreds, even thousands, of miles an hour. This is because the gravity forces—called "g's"—would instantly tear a human-made aircraft apart, killing both the pilot and passengers.

How many g's can a UFO withstand? UFOs have been reported easily pulling 20 g's. And yet, scientific studies show that humans will pass out at five or six g's, and they will die at a mere nine g's.

According to **ufologists** and other highly trained observers, one of the most amazing UFO traits is their ability to move forward and pause without any sign of acceleration or deceleration. They can simply go from either a complete stop to full speed, or from full speed to a complete stop *instantly*—without speeding up or slowing down.

UFOs possess other incredible characteristics as well, such as being able to hover motionless, even in Earth's most powerful windstorms. Many have been spotted doing just this, much to the amazement of the eyewitnesses.

UFOs also have the ability to appear from nowhere, as well as disappear into thin air at the drop of a hat. We have many reports of unknown objects "turning on" and "turning off," kind of like a light switch, so that one second they're right in front of your eyes and the next second they're gone.

My own theory is that some alien species have managed to learn how to operate their craft in regions of the light spectrum that we humans can't see in, such as infrared and ultraviolet. If true, there could be UFOs all around us nearly all the time and we wouldn't be aware of them.

HOW DO UFOS WORK?

We have no idea how UFOs manage to perform these incredible maneuvers, maneuvers that violate every one of the laws of physics as we currently know them.

> UFO FACTOID
> Most of the contrails you see in the sky are made by jets. But sometimes UFOs make them as well.

Some think that UFOs may be powered by some type of anti-matter system. Others believe that they run off anti-gravity machines, or even some kind of "gravity amplifier," both which would use large magnets. Using magnets for propulsion creates quite a bit of instability, which would help to explain the tendency that some UFOs have for wobbling, trembling, and hopping.

But such means of propulsion are so far beyond our current aerospace technology that we can't even begin to imagine how UFOs use them to fly, let alone how to build one.

One clue may come from the fact that some UFOs leave vapor or exhaust trails behind them called "contrails," while others emit huge tails of hot steam, sparks, flames, and smoke. This makes us think that UFOs may be powered by engines similar to our own conventional aircraft. But what about the thousands of UFOs that don't give off any vapor, exhaust, steam, fire, or smoke?

The riddle of how they fly is made all the more mysterious by the fact that on most UFOs there's no visible engine or propulsion system, no intake ports or exhaust openings, and in nearly all cases, no wings or appendages (used for lift and stabilization) of any kind. And yet not only can UFOs fly, climb, and dive just like our own aircraft, they can also start and stop instantly, shoot straight up or down, hover motionless (even in high winds), and turn sharp corners (even at high speeds)—maneuvers that are impossible for our vehicles.

WHERE DO UFOS COME FROM?

Since UFOs are real it's obvious that they originate somewhere. But where? Like many questions having to do with UFOs, we don't yet know the answer. This hasn't stopped us from trying to figure it out, however.

> UFO FACTOID
> The behavior of UFOs can be baffling. In 1974 a glowing UFO appeared over France, shooting out four beams of light that stopped in midair, a feat that goes beyond all known scientific laws. It made a number of abrupt ninety-degree turns, then vanished.

Most people seem to think that UFOs come from nearby stars or constellations, such as Sirius, Orion, Lyra, Draco, Andromeda, Pleiades, or Arcturus. Others believe that the home of most UFOs is on Alpha Centauri, Earth's nearest stellar neighbor at only four light-years away. Still others think that UFOs originate right on our own planet, and that they come from underground, or even underwater somewhere, as the existence of USOs seems to indicate.

One of the more realistic theories of where UFOs come from has to do with the fact that, as Albert Einstein proved, gravity can alter time. If UFOs *are* powered by anti-gravity devices, then it's possible that they may be able to bend or distort time in some way. And if they can distort time, the vast distances of outer space wouldn't matter: UFOs could come from, and travel to, just about anywhere in the Universe in an instant.

Here's another hypothesis. Whoever or whatever is behind UFOs is obviously technologically superior to we humans. If, for instance, aliens are 1 million years older than us, then they're certainly very highly evolved creatures who know things about the laws of physics that we haven't had time to learn. They would undoubtedly be able to travel instantaneously from parallel universes to our universe, or ride through wormholes—which bypass time and space—like the people and aliens in *Star Trek* and *Star Wars*.

UFO FACTOID
Are aliens and UFOs visiting us from our own future? Many think so.

One more bizarre possibility: UFOs may be visiting us from our own future. In other words, the aliens who are traveling around the Earth right now may actually be us—only hundreds of thousands of years from now. How could this be possible?

Again the answer lies with Einstein, who proved that if you can travel faster than the speed of light (186,000 miles per second), then you can travel back in time. Though many mainstream scientists believe that traveling at this speed is impossible, there's no reason why aliens—who are far more scientifically advanced than we are—couldn't have mastered the barriers of time and space.

If this is true then we may be embryonic aliens, baby aliens that haven't had time to fully grow and develop yet. As shocking as this sounds, this would make aliens fully evolved humans!

UFO FACTOID
Although tens of thousands of UFO sightings have been logged during the modern UFO era, the CE1 remains the most common.

HOW UFO SIGHTINGS ARE ARRANGED

If you saw the movie *Close Encounters of the Third Kind*, then you already know a little bit about how UFO sightings are categorized. But you might have always wondered what the "first" and "second" kinds of "close encounters" are. Actually, there are five different classifications of UFO interactions.

The first is called a "Close Encounter of the First Kind," or a CE1, for short. Here someone sees a UFO within 500 feet, but there isn't any kind of contact between the observer(s) and the UFO. It's just a sighting.

In a "Close Encounter of the Second Kind," or CE2, a UFO interacts with the environment in some way, leaving behind evidence of its presence. This would include leaving depressions or imprints in the ground, disturbing electronic equipment (such as TVs, radios, cameras, computers, cell phones, lights), or causing physical effects on living things (such as burns, cuts, nausea, or illnesses).

In a "Close Encounter of the Third Kind," or CE3, people not only see a UFO, but they also see the occupant(s) inside the UFO as well. These creatures are what we call "aliens," or ETs (short for "extraterrestrials"). Usually aliens are **humanoid**—that is, they look something like a human—in appearance, though not always, as we will soon see.

In a "Close Encounter of the Fourth Kind," or CE4, humans interact in some way with aliens, such as by talking. More typically the humans are "abducted" by the aliens and taken on board their spacecraft.

In a "Close Encounter of the Fifth Kind," or CE5, humans see or make contact with UFOs by actively seeking them out. It has been reported many times that UFOs can be lured into showing themselves if people flash bright lights, like flashlights or laser pens, up toward the sky at night. Sometimes UFOs even flash their lights back—in the same pattern.

UFO AT RENDLESHAM FOREST

We've learned many fascinating facts about what ufologists call the "high strangeness" world of UFOs. But what would it be like to really see one?

To find out, let's take a look at some more casebook studies, all gleaned from the files of actual UFO reports. British ufologist Timothy Good calls the following incident "one of the most sensational UFO events ever reported by military personnel."

UFO FACTOID
On a cool night in December 1980 a mysterious triangular-shaped object with red lights was seen flying slowly between the trees in England's Rendlesham Forest, near the Bentwaters-Woodbridge military base. Witnessed by several high-ranking officers, three perfect depressions were discovered in the ground, along with high levels of radiation in the trees.

Casebook Study 5: The time was midnight. The day was December 30, 1980. The place was **RAF-USAF** Bentwaters-Woodbridge, a joint British-American military base near Ipswich, Suffolk, England, about 100 miles from London.

Several soldiers were on watch that night when they saw what looked like a fire in a woods nearby called Rendlesham Forest. After reporting the strange scene to their commanding officers, they were told to go and check the situation out.

As they got closer to the woods their jeep engine mysteriously stopped running, so they got out and began to walk the rest of the way in. They couldn't help but notice then that hundreds of animals were running wildly from out of the woods, while at the same time they could hear animals from a nearby farm going "into a frenzy."

A few minutes later the men came upon three deep circular impressions (like the feet of a large tripod) spaced equidistant apart and burned into the ground. Using a Geiger counter, readings of the holes (and of nearby trees) showed high levels of radiation, something completely out of place in a forest.

As the men walked further into the dark woods, a bizarre "pulsing," triangular-shaped object came into view. It was maneuvering slowly through the trees, pausing, hovering, then moving up and down.

The soldiers then noticed a yellow fog or mist beginning to appear, hanging in the air about three feet off the ground. From out of it a new craft emerged. The men would later describe it as an unusual round object, similar to a "transparent aspirin tablet," at least fifty feet wide, with numerous colored lights on top and blue lights on the underside.

UFO FACTOID
An English newspaper headlined the Rendlesham Forest incident, the most sensational of all British UFO incidents, this way: "UFO LANDS IN SUFFOLK!"

By now large groups of other soldiers were arriving and helicopters were buzzing around overhead. Some soldiers had set up movie cameras and were filming, while security officers were forming a large human circle around the object.

Two cows from a neighboring farm, strangely captivated by the curious craft, wandered over and stared at it from only a few feet away. Some of the soldiers, unable to accept what they were seeing, broke down and cried and had to be taken away and treated for shock by military medics.

Just then, to everyone's astonishment, a bright red light came soaring in from the sky overhead and floated down through the trees until it was hovering silently over the huge "aspirin." The red light then "started dripping like molten metal," after which it suddenly exploded into thousands of small colored lights. But the weirdness was only just beginning.

When these sparks faded away the men realized that the red light and the "aspirin" had completely disappeared. In their place sat a huge black disk with a dome on it. The soldiers on the far side of the disk saw strange alien beings inside of it, and even communicated with them.

At this point everyone's attention was drawn to the sky, where they saw three star-like objects with red, blue, and green lights, darting around, performing sharp angular maneuvers. Out of one of the objects came a brilliant green laser-like beam of light, about ten inches in diameter, that fell on the ground right in front of the soldiers' feet.

Then abruptly, all of the mysterious crafts disappeared, flying straight up into the night sky.

The entire event, later called "the Bentwaters-Woodbridge UFO Incident," had been photographed, filmed, captured on radar tapes, and seen by not only dozens of highly trained Air Force personnel, but also by numerous civilians living in the surrounding area. The day after the event all of the witnesses filled out detailed UFO sighting reports, testimonies that they solemnly stand by to this day.

UFO FACTOID
The Bentwaters-Woodbridge affair was witnessed by countless military officers and soldiers, *and* recorded on film, video, and radar. Yet both the U.S. and the British governments pretended that it never took place.

THE HALT TAPE

One of these witnesses was the Deputy Base Commander at the time, **USAF** Colonel Charles Halt, who made an on-site tape recording while he was in the forest pursuing the UFO. This tape, ruled "authentic" by the British government, has been broadcast on international TV and is now in the public domain. A book called *Left at East Gate*, by Larry Warren and Peter Robbins, examines the affair.

What follows are a few portions of Colonel Halt's actual tape as he and his men approached the UFO in Rendlesham Forest, recorded on the night of December 27-28, 1980. After receiving several strong clicks on their Geiger counter, they discovered an "eerie heat source" on some nearby pine trees, along with "freshly broken pine branches" on the ground "from some fifteen to twenty feet up."

They also came upon "strange abrasions" on the bark about three feet off the ground and an "opening" far up in the treetops. Farm animals nearby now began making "very strange sounds." At this point the UFO came into view.

HALT: You just saw a light? Where? Wait a minute. Slow down. Where?
SOLDIER: Right on this position here. Straight ahead, in between the trees. There it is again. Watch, straight ahead, off my flashlight there, sir. There it is. . . .
HALT: Oh yeah, I see it too. What is it?
SOLIDER: We don't know, sir. . . .
HALT: There's no doubt about it. There is some type of strange flashin' red light ahead.
SOLDIER: Sir, it's yellow.
HALT: I saw a yellow tinge in it too. Weird! It appears to be maybe movin' a little bit this way? It's brighter than it has been.
HALT: It's comin' this way. It is definitely coming this way! Pieces of it are shooting off.
SOLDIER: At eleven o'clock.
HALT: There is no doubt about it. This is weird!
SOLDIER: To the left . . . two lights! . . .
HALT: Keep your flashlights off. There's something very very strange . . .
HALT: Pieces are fallin' off it again! Strange! Whew!
HALT: Let's push to the edge of the woods up there. Can we do it without lights? Let's do it carefully, c'mon!
HALT: Okay, we're lookin' at the thing. We're probably about 2 to 300 yards away, and it looks like an eye winkin' at ya, it's kinda movin' from side to side. And when you put the Starscope on it, it's like it has a hollow center, a dark center, yeah, like the pupil of an eye lookin'at ya, winkin'. And the flash is so bright to the Starscope that it almost burns your eye. . . .
HALT: . . . we now have multiple sightings of up to five lights with a similar shape and all. . .
HALT: . . . we're seeing strange lights in the sky. . . .
HALT: We see strange strobe-like flashes to the . . . rather sporadic, but there's definitely something there. Some kind of phenomenon.

HALT: At about ten degrees horizon directly north, we've got two strange objects, ah, half moon shape, dancin' about, with colored lights on 'em. . . .

HALT: . . . they're both heading north. Hey, here he comes from the south; he's coming toward us now!

HALT: Now we're observing what appears to be a beam comin' down to the ground.

SOLDIER: Look at the colors!

HALT: This is unreal!

HALT: 3:30 [A.M.] and the objects are still in the sky, although the one to the south looks like it's losin' a little bit of altitude. We're turning around and headin' back toward the base. The object to the south is still beaming down lights to the ground.

HALT: One object still hoverin' over Woodbridge base at about five to ten degrees off the horizon. Still movin' erratic and similar lights beaming down as earlier. [End of tape.]

As might be expected, neither the U.S. government or the British government ever openly discussed the incident with the public. Colonel Halt was never even debriefed, and to this day not one military official has ever asked him about the experience.

"Britain's Roswell" was covered up and buried as quickly as possible by the authorities, and in the process the three landing gear depressions Halt found became "birds nests" and "rabbit burrows," while the UFO he witnessed became a "lighthouse." The base was deactivated in 1993, and most everyone above the rank of colonel acted as if nothing out of the ordinary ever happened at RAF-USAF Bentwaters-Woodbridge in late December 1980.

CHAPTER 1 AT A GLANCE

- A UFO can be seen anywhere, at anytime, by anyone.
- Not every unidentifiable flying object is a UFO: 90 percent are later found to be known objects.
- At least 10 percent of all sightings of unidentified flying objects turn out to be genuine UFOs.

- UFOs can be divided into two basic categories: those that move in the air, UFOS, and those that move underwater, USOs.
- UFO sightings are classified into three visual categories: DDs, NLs, and RVs.
- UFOs can be any size, from a small dot of light to a super UFO many miles wide.
- A UFO can be any shape, but most look something like a saucer or a triangle.
- Many UFOs can shape-shift, going slowly or rapidly from one form to another.
- A UFO can be any color, from black to rainbow-colored; but most seem to be bright silver or white.
- Most UFOs have navigation lights of some kind, that either pulsate, flash, or rotate. These lights may flash off and on in some kind of set pattern, or completely randomly and independently. UFOs can also be entirely lightless, or they may have powerful searchlights that explore the ground. Some display short, abruptly ending beams of light, a technological feat far beyond our scientific capacity and understanding.
- UFOs may have one window, many windows, or no windows. The windows themselves may be small like a porthole, or large like a picture window.
- UFOs make many different types of sounds, from whines and whistles, to beeps and jet-like sounds. Most UFOs, however, are completely silent.
- UFOs can be seen flying alone or in huge formations with dozens, sometimes even hundreds, of other UFOs.
- USOs, or unidentified submerged objects, have been reported for thousands of years, dating back to ancient times. These uncanny craft display a bewildering set of abilities: they can dive to incredible depths and travel at velocities close to the speed of sound, and most are capable of taking to the air and even shooting up into space. All of this is far beyond our technology.

- A UFO can hover, float, hop, weave, wobble, jump, swing back and forth, dance, dart around, dive and climb, reverse, land on the ground, or shoot straight up in the air.
- UFOs can not only travel faster than any human-made aircraft, they can also completely outmaneuver any human-made aircraft. In fact, UFOs can move much faster than the fastest bullet or missile: many have been clocked at speeds of between 60,000 and 100,000 mph. According to one government official they may even be able to travel at the speed of light: 669,500,000 mph, something Einstein said was impossible.
- A UFO can make itself appear or disappear at will, simply by turning itself "on and off." It can also defy every known law of physics, including turning a ninety-degree corner at full speed without slowing down.
- We don't know how UFOs work, but they seem to use Earth's gravity and magnets in some way.
- No one knows for sure where UFOs come from, but most likely they originate somewhere in outer space, probably outside our solar system.
- Some people speculate that some UFOs and aliens come from under our seas, where they've constructed huge underwater bases.
- The aliens who build and control UFOs may know how to go back in time, travel through wormholes, or pass through parallel universes.
- Interactions with UFOs are arranged in five categories: CE1, CE2, CE3, CE4, and CE5.
- Ufologists are people who study UFOs.
- The Bentwaters-Woodbridge UFO incident is the most famous UFO case to date in Great Britain.

ALIENS: MYSTERIOUS INTRUDERS FROM SPACE

"These objects [UFOs] are conceived and directed by intelligent beings of a very high order. They do not originate in our solar system, perhaps not in our galaxy." – Dr. Herman Oberth, one of the founding fathers of astronautics, 1954

ALIENS: MORE QUESTIONS THAN ANSWERS

We've learned all about UFOs. But there are still many other things we need to know. Who's flying them? Where do they come from? And what are they doing here?

These are excellent questions. Unfortunately, when it comes to aliens we still have more questions than answers. However, it's obvious that some kind of intelligent lifeform is behind UFOs.

As nuclear physicist Bob Lazar says, since UFOs are real, they must be built by some kind of intelligent creature, and that intelligent creature must be an alien. To build UFOs the aliens must have a factory, and to have a factory there must be lots of aliens. And if there are lots of aliens there must be an entire alien civilization somewhere.

> UFO FACTOID
> In the 1980s scientist Bob Lazar is said to have worked on alien spacecraft at a secret Nevada military base called S-4.

WHAT'S AN ALIEN?

Let's begin our exploration of the alien civilization by defining what an alien is.

The word alien of course has many meanings. It can mean a person who's from a different family, race, or country than you.

But the word can have another meaning: "a being, a creature, or a lifeform that isn't from planet Earth." This definition for the word alien, what ufologists technically call an EBE (for "extraterrestrial biological entity"), is the one we'll be using throughout this book.

HOW MANY KINDS OF ALIENS ARE THERE & WHAT DO THEY LOOK LIKE?

There seem to be almost as many types of aliens as the UFOs they fly in. We know this because aliens have been described as everything from blobs of jelly to robot-like creatures, from bizarre looking insects to beings that look like vegetables. Some people have even seen aliens who look exactly like humans.

THE EIGHT BASIC CATEGORIES OF ALIENS: FOUR-FINGERED & FRAGILE: THE GRAYS

Despite the wide variety of aliens (currently, at least sixty reported different species), most seem to fall into one of eight basic categories:

> UFO FACTOID
> Grays are the classic alien, appearing in countless films and TV shows. Skeptics say that the so-called "eyewitnesses" of Grays are using these Hollywood images to describe them. People however, were seeing Grays long before the invention of movies, or even electricity.

1. The Gray: This is the most widely sighted type of alien, and the type you're most likely to see in a science fiction movie or TV show. The Gray has a large head that looks something like a light bulb, with huge black eyes that curve around its head, like wrap-around sunglasses.

It's usually short, about three or four feet in height, and has a fragile-looking body, with a thin neck, thin torso, and thin arms and legs. These characteristics make the Gray look a bit like a newborn human baby. Eyewitnesses who have seen Grays close up say that they only have four fingers on each hand (actually, three fingers and a thumb) and four toes on each foot.

The role of the Grays seems to be to maintain their spacecraft, or what we call UFOs. They may be assistants or underlings, individuals of lower rank who follow the orders of their superiors and work behind the scenes.

UFO FACTOID
Grays are the most commonly seen type of alien and may be related to humans.

The Grays have a sinister quality about them and are extremely frightening to those who've seen them. And yet, for some reason we don't fully understand, people who've had encounters with the Grays say that not only do they feel a deep love for these infant-like beings, they also feel deeply loved by them.

Perhaps this is because, as some have suggested, we're somehow related to the Grays. One theory is that the Grays created humans by mixing their genes with those of a prehistoric humanoid (such as *Homo erectus*), creating what we call "modern humans," or *Homo sapiens sapiens* (Latin for "wise wise man"), the hopeful name we've given to ourselves.

Another theory that we've looked at states that the Grays are really us in the future, many thousands of years from now. According to this belief the Grays have come back in time in order to try and help save humanity from itself.

Either way, if any of this is true, we and the Grays are cousins.

THE BLOND, THE BOLD, & THE BEAUTIFUL: THE NORDICS

2. The Nordic: This is the next most commonly seen type of alien. The Nordic gets its name from the fact that it looks somewhat similar to someone from Norway: they're tall, and they have blond hair, blue eyes, and white skin, just like typical human Norwegians.

UFO FACTOID
Like the Grays, the Nordics are also believed to be related to humans in some way.

Humans who have come into contact with Nordics find them to be friendly, kind, and nonthreatening. Some even consider them to be "beautiful" (at least by human standards), for it's said that they have a charismatic "movie star" quality about them.

But there is one major difference between an alien Nordic and a human Nordic: the alien Nordic is self-luminescent; that is, its entire body glows from within with a strange light, including its hair!

Nordic aliens seem to be "in charge," bossing other types of aliens, like the Grays, around. This "chain of command"—with the Nordics acting as supervisors and the Grays acting as subordinate assistants—makes us think that the world of the aliens is more like a large military federation of different alien species (as on *Star Trek*) than a simple community of civilians.

STICK-LIKE & TERRIFYING: THE INSECTILES

> UFO FACTOID
> Eyewitnesses say that the Insectile alien looks very much like a giant praying mantis.

3. The Insectile: This class of aliens is not seen very often, but when humans do encounter them they find them extremely terrifying. Why? Because with their enormous oblong heads, huge cold black eyes, and stick-like bodies, they resemble giant insects—praying mantises to be exact.

Some think that the Insectiles are specially made to pilot UFOs. Why?

If a woman or a man flew in a UFO, their body would be torn apart by the tremendous gravitational forces, or "g's," created by the craft. But the Insectiles are actually well-suited to such forces, since they have bony but very strong limbs, making them more resistant to the abrupt changes in gravity that a UFO encounters.

The Insectile also seems to be the "doctor" of the alien world, performing surgery and other procedures, and caring for the aliens' medical equipment. Also, unlike the friendly Nordics who seem to like us, the Insectiles have a cold and menacing presence and actually seem to despise us. In fact, they treat humans as callously as we humans treat insects. No wonder people are afraid of them.

PEACEFUL & SPIRITUAL: THE ASIANS

> UFO FACTOID
> Very little is known about the Asian alien because very few people have encountered them.

4. The Asian: This type of alien is rarely seen by humans, but it's common enough to be listed as one of the eight basic categories.

Like the Nordic aliens the Asian aliens get their name from the fact that they closely resemble a type of human: people from Asia. Those who've seen them say that they're quiet and pleasant, and that they seem to be religious or spiritual in some way. Other than that we know almost nothing about them.

ALMOST . . . BUT NOT QUITE HUMAN: THE ROSWELLIANS

UFO FACTOID
The U.S. military has performed secret medical operations on the bodies of dead Roswellians in an effort to find out more about them. Some people claim that our government houses several live Roswellians, captured during UFO crashes.

5. The Roswellian: This is a humanoid alien that looks eerily similar to a person, but is not quite human. It has a bald head and dark piercing eyes, both a bit bigger than ours. It's a frail creature that stands about four feet tall. It's said to have four fingers on each hand and four toes on each foot. The Roswellians are quite harmless and nonthreatening to humans and actually seem to be afraid of us.

They get their name from the town of Roswell, New Mexico, where, as we'll learn in our next chapter, one of their vehicles plummeted to Earth in 1947. The craft, with its three alien occupants, was recovered by the U.S. military and examined. Later an autopsy was performed on the bodies of the dead aliens and a film said to be the actual footage of these operations was shown on international TV.

FETUS-LIKE & MYSTERIOUS: THE NEONATES

6. The Neonate: Not much is known about this species of alien, just their appearance. Even when full grown they look like a one-month-old human baby, or to be more accurate, like a human fetus: a child still growing inside its mother's womb. This is why they're called the "Neonates." Neonate comes from the Latin words *neo*, meaning "new," and *natalis*, meaning "at birth." In other words, a baby.

HALF-ALIEN, HALF-HUMAN: THE HYBRIDS

7. The Hybrid: This creature is one of the more unusual extraterrestrials, because it's half-alien and half-human. This is why they're called "Hybrids."

UFO FACTOID
The Hybrids are the result of the aliens cross-breeding their DNA with our DNA. Many human women claim to have been implanted with the fetuses of these experiments. After giving birth the hybrid babies are forcibly taken from the women, an emotional trauma from which they never fully recover.

The word hybrid comes from the Latin word *hybrida*, which means "something that's a blend of two things." Some people feel that aliens are kidnaping humans, taking their DNA (or genetic material) and mixing it with their own, and that this is how the Hybrids are created.

Why are the aliens creating a creature that's a mixture of alien and human? It's thought that this is a way for them to prevent their species from completely dying out.

It seems that their home planet may be deteriorating, burning up as their Sun expands (our Sun too will expand into what's called a "red giant," reaching maximum size in about 7.6 billion years, by which time all of the inner planets in our solar system will be consumed by heat and fire, including the Earth). If true, they're looking for a new world to occupy, and ours is the nearest, or perhaps the most liveable, world they're aware of.

But they can't survive here as they are. Creating this new species, the Hybrids, would make it easier for their kind to live on our planet, allowing them to preserve their DNA for a possible future return to a genetically engineered, fully alien being.

UFO FACTOID
According to eyewitnesses the Reptilians are arguably the scariest of all the alien species.

SCARY & SCALY: THE REPTILIANS

8. The Reptilians: These aliens, Reptoids as they're also called, are said to be the most sinister of all. And this is not surprising. They have cat-like eyes, scales and claws, and look something like a cross between an alligator, a lizard, and a dinosaur.

Standing eight to nine feet tall the Reptilians are quite intimidating, and not just because of their huge dinosaur-like appearance, but because of their attitude toward humans. They consider us a vastly inferior species, perhaps the way we think of rats.

Luckily for us sightings of Reptilians are the rarest of all alien types.

HOW DO WE KNOW IF ALIENS ARE REAL?

Aliens, like the UFOs they pilot, seem so improbable that many scientists and skeptics—even those who have seen UFOs—can't accept the idea that they're real living beings. And yet we know they're real because millions of people from all over the world have seen them, touched them, smelled them, and heard them.

People who see aliens, or come into contact with them in some way, are called "experiencers." Experiencers are usually ordinary everyday people who come from every background, age group, and country. From children to senior citizens, from soldiers to dentists, from nuns and priests to road repair crews, from postal employees to fire fighters, all these and more have seen aliens.

UFO FACTOID
Contrary to the claims of nonbelievers and skeptics, experiencers are ordinary people who happen to experience something extraordinary.

As they do with UFOs, skeptics and debunkers (people who don't believe in UFOs) would like to use the three "h's" (hoaxes, hysteria, and hallucinations) to discount the testimony of experiencers. But this is impossible when we consider three things:

1) Experiencers don't ask to see aliens or to interact with them. Aliens nearly always come into a person's life uninvited, and usually unwanted.

2) Very few, if any, experiencers ever seek publicity, fame, or money because of what's happened to them. The truth is that most experiencers are sane, private, sincere, serious-minded, reliable people who just want to be left alone. Indeed, this is why the majority of experiencers never tell anyone about their encounters with aliens, not even their own family members.

3) The eyewitness testimony of an experiencer would stand up in any court in the world. So why isn't it good enough to prove the reality of aliens?

The truth is that most scientists don't want to have deal with something they can't understand, and which would overthrow every known scientific theory and law. It's easier to ignore the idea of aliens and pretend that people who see them are just imagining things.

After years of study as the **USAF**'s chief scientific consultant on UFOs, however, one famous American astronomer and skeptic, J. Allen

Hynek, became a believer. There are simply too many highly intelligent, highly educated, highly trained individuals reporting UFOs, Hynek stated, for it all to be misidentification and imagination.

And he was right.

UFO FACTOID
Known aliens range in size from just a few inches to over seven feet tall.

UFO FACTOID
Prehistoric people believed in a Mother-Goddess rather than a Father-God, the Supreme Being most worship today. Is this great female deity the same one who leads the aliens? A belief in the Mother-Goddess lasted well into ancient times here on Earth, as mythology clearly reveals. Her various names are well-known to us even now: Juno, Venus, Aphrodite, Kelle, Cybele, Asherah, Isis, Astarte, Mari/Meri, Kali, Sunna, Atthar, Freya, and Aditi, to name a few. Our weekday Friday takes its name from a Scandinavian version of the Great Mother-Goddess known as Frigga. Catholics still worship her as the Virgin Mary, while Mormons venerate her as the "Heavenly Mother."

WHAT SIZE ARE ALIENS?

Some aliens are truly tiny, such as the one witnessed in 1970 by a group of Malaysian schoolboys: it was a mere three inches tall. But others have been seen that are much bigger, such as the alien spotted by someone in 1968 in West Virginia: the carrot-like creature stood some seven feet tall from head to toe.

It's entirely possible that some may be as small as microbes and others as big as houses, but the two examples cited above are the currently known extremes. In short, while some types of aliens are actually just about the same height and size as we humans, the majority seem to be shorter and smaller than an adult, averaging around four feet in height.

HOW LONG DO ALIENS LIVE?

Because our knowledge of aliens is so limited, we really don't know the answer to this question. Therefore it's possible that some alien species may live for as little as a year or two while others may live for thousands of years.

We do have a clue about one particular type of alien though. A scientist who had the privilege of doing an autopsy on a captured, dead humanoid alien estimated that this particular individual, a female, was at least 200 hundred years old.

SECRETIVE & SUPERIOR: THE NATURE OF ALIENS

The aliens are in total control of when, how, and where they appear to humans here on Earth. And we only know what they allow us to know about them, which is why we know so little. Thus they continue to remain truly baffling creatures.

But we can get an idea about the nature of aliens from the stories of the many experiencers who've come into contact with them. Here is what we've learned.

Aliens are incredibly secretive, powerful beings who come from some unknown region in space. They're also highly evolved, and may be as much as 1 billion years older than us. We modern humans (*Homo sapiens sapiens*) are, after all, only 195,000 years old, while the Universe is around 13 billion years old. This huge time span would give other lifeforms in the Universe a chance to evolve far beyond us.

Some reports lead us to believe that the aliens operate under an all-powerful female leader who has assigned each of them certain roles and tasks. This female seems to be similar to the great Mother-Goddess that was worshiped in ancient times here on Earth, and who is still venerated by millions of people all around the world as the "Supreme Being." Experiencers tell us that this all-powerful female is called "the One" by the aliens.

UFO FACTOID

Our connection to aliens is far more profound than most people realize. It seems that not only did they help humans develop culture (e.g., writing and agriculture), but, according to some, they may have created our species itself.

Because evidence suggests that they've been visiting Earth for millions of years, some people think that, as superior beings, the aliens might have helped us evolve from prehistoric humans to modern humans. How did they do this?

Probably by assisting us in learning how to use fire and the wheel, how to make clothing, how to practice agriculture, how to develop writing, and how to construct buildings and make cities. The ubiquitousness of ancient, complex monumental pyramids around the world (whose bases, shapes, and heights are often of the exact same dimensions), seems to lend credence to this idea.

Others think that the aliens, especially the Grays, actually created us by combining their DNA with the DNA of primitive ape-like creatures. Maybe this is why, in some ways, we look so similar to both the Grays and to nonhuman primates.

If this theory is true it would also help answer a problem that has perplexed paleoanthropologists for decades: there is a long and mysterious gap in the fossil record between primitive humans and modern humans. How could modern humans appear so suddenly, and fully formed, in the fossil record with almost no signs of having evolved from earlier more primitive types of humans?

> UFO FACTOID
> Aliens seem capable of communicating in a multitude of ways. They can even speak to a human in his or her own native language. The most common method of alien communication, however, is mental telepathy.

If we're actually related to the aliens in some way then this also gives us a clue as to why they're visiting Earth, because this would mean that we're cousins with them. Like any family member they must be concerned about us, about our future, about how we're treating Mother Earth.

In fact, when experiencers have asked the aliens what their purpose here is, the aliens tell them that they have a special "mission," one that involves saving the Earth from pollution, overpopulation, and nuclear war. Aliens have also told experiencers that they're "from the Spirit," that they "love humans," that, in other words, they're visiting Earth to help keep humanity from destroying itself.

HOW DO ALIENS COMMUNICATE?

Although aliens do have small mouths they don't usually use them to communicate the way we do. Reports suggest that they don't normally speak at all. But when they do almost anything can come out.

For instance, aliens have been heard speaking with very high squeaky voices, with very deep low voices, and with weird electronic-sounding voices. And yet some say that they make "strange gurgling sounds deep in their throats" when they talk.

Aliens also seem to be able to speak in any human language, which is why they're understood by experiencers all over the world. But for some unknown reason many experiencers say that the aliens speak to them in the Gaelic tongue (an ancient Irish, Welsh, and Scottish language). Others say they talk in what sounds like Russian or Chinese. Obviously each alien race must have its own language.

> UFO FACTOID
> The vast majority of reports suggest that aliens, particularly the Grays, are vegetarians. Not only have they been seen ingesting a strange green fluid, but autopsies performed on them by surgeons reveal that they have few or even no teeth.

Despite these testimonies, however, it's actually quite rare to hear aliens talking out loud. More typically, according to thousands of eyewitnesses, aliens communicate telepathically; that is, they speak without words, directly from one mind to another. So if an alien were to speak to you, its mouth wouldn't move. Instead you'd hear its voice inside of your head, speaking as clearly as if it were standing right next to you.

Some experiencers have reported that certain aliens, like the Grays, communicate through their huge black eyes. How they do this we don't know, but oddly they seem to have the power to calm people down and make them feel loved just by looking at them.

WHAT DO ALIENS SMELL LIKE?

Most experiencers who've been in the presence of aliens don't report any scent at all. But the few who have smelled them say that they usually have a strong, harsh, and often bitter aroma. Descriptions range from the smells of wet cardboard, ammonia, wood, and cinnamon, to the smells of cheese and even decaying or rotten meat.

WHAT DO ALIENS FEEL LIKE?

Very few experiencers who have actually touched an alien can remember what one feels like. This is probably because aliens usually block a person's memory of the experience. But the few who do remember say that their skin is very soft to the touch, almost like silky plastic. As such, it may be something similar to the skin of a dolphin.

WHAT DO ALIENS EAT?

Another one of the many mysteries surrounding aliens is what they use for food. They appear to require very little in the way of sustenance. But the truth is that few people have ever seen an alien eat, so we can't be sure.

One person who did reported that she saw an alien drink a thick green liquid. Outside this one incident, we have very little information on the aliens' diet—although many believe that most species seem to be vegetarians.

THE AMAZING POWERS OF ALIENS

Along with their ability to communicate telepathically and live on green liquid, aliens have many other astonishing abilities. The most startling one is that they can move through solid objects, like closed doors and windows, or through ceilings and walls. They have also been seen floating over floors instead of walking on them.

How do they do this? We don't know. According to the laws of physics—as we understand them—these things are impossible. But they're not impossible for the aliens. They must have knowledge about the way the Universe works that we haven't even begun to think about yet.

We can get a better understanding of the aliens' mysterious abilities by looking at an actual case of contact between aliens and humans.

A REAL LIFE CLOSE ENCOUNTER WITH ALIENS

One of the best documented close encounters took place in 1985 in upstate New York.

> UFO FACTOID
> Whitley Strieber's experiences with UFOs and aliens are among the most detailed and well documented in the annals of ufology.

Casebook Study 6: On the night of December 26, author Whitley Strieber, his wife Anne, their son, and two guests, were sleeping peacefully in Whitley's country cabin. Suddenly he was startled awake by a loud swooshing sound coming from his living room.

As he sat up in bed he noticed a small dark figure peering around the door to his bedroom. It had two large black eyes, which were looking directly at him. Terrified, he started to reach for the light next to his bed, but found that he was paralyzed and couldn't move.

Before he knew it the strange creature came rushing over to the bed, where it stood staring blankly into his eyes.

Whitley pinched himself to see if he was dreaming. He wasn't. He was wide awake. He could even hear his heart pounding. As he looked over at his wife, she was sleeping soundly next to him. He wanted to yell to wake her, but his mouth wouldn't open. He sensed that he was under the complete control of some mysterious power.

Then Whitley felt his whole body lift up off the bed. As he looked down he saw his bed moving away from him. But it wasn't his bed, it was him: he was moving slowly through the air. Using some kind of strange force, the little creature then levitated him out of the bedroom, out of his house, and out into the cool winter night. Then everything went black.

UFO FACTOID
Like many other believers, Strieber started out as a hard-boiled skeptic. This disproves the theory that experiencers only see UFOs or interact with aliens because they believed in them beforehand, as non-believers claim.

When he came to, Whitley found himself laying on a table in the middle of a small circular room. Though he wasn't tied up, he still couldn't move his body, only his eyes. As he peered around the room, he saw a group of Grays gathering around him. They seemed to be busy preparing him for some kind of surgery.

Before he could resist they began to perform a series of odd, horrifying, and painful medical tests on him. Screaming in agony and fear, Whitley passed out again.

The next thing he remembers is waking up in his own bed the following morning. He had a dim memory that something terrible had happened during the night, but he couldn't recall exactly what it was. Was it a nightmare or just his imagination?

Later, under hypnotic regression, he relived the events of that frightening evening and realized for the first time that he had encountered a group of beings that were not from this world. "Impossible!," he thought.

Whitley had always loved science, so when it came to "UFOs and aliens" he was a complete skeptic. This made the experience difficult to accept. And yet it seemed so real. After all, he had been wide awake through almost the entire event.

Starting to doubt his sanity Whitley had himself tested for various mental and physical diseases. But all of the tests came back normal: he was perfectly healthy. He even took three lie-detector tests—and passed all three.

It was at this point that Whitley discovered that the two guests who stayed in his cabin that night also had several strange experiences. Comparing notes he found that during the exact time he was being taken by the aliens, the two of them had woken up and seen weird lights, heard unusual sounds, and felt bizarre physical sensations. He could no longer deny the reality of what had happened to him.

The experience changed Whitley Strieber's life forever, and from that day forward he became an impassioned believer in UFOs and aliens. He's written numerous books about his encounter with the aliens, all which have been read by millions. One of them, called *Communion*, was later made into a popular movie.

THE MOST OFTEN ASKED QUESTION ABOUT ALIEN BEINGS

Whenever the topic of alien beings comes up, someone (usually a skeptic) will always ask the following question: "If aliens are real, why don't they just land on the White House lawn and openly introduce themselves to the world?"

The simple answer is that they're afraid to. But why?

Since aliens have been observing us for thousands of years it must be glaringly apparent to them by now that we're a very aggressive species, and that because of this we're far more apt to make war than peace.

The aliens aren't imagining this: every nation on Earth that can afford one has a military, complete with ships, jets, tanks, guns, ammunition, bombs, and soldiers. At any one time there are at least sixty wars going on in the world, somewhere, someplace. Often there are many more. The history of nearly all nations has been written in blood.

> UFO FACTOID
> It's more than obvious why aliens don't simply land on the White House lawn and introduce themselves: they're afraid of our aggressive and often violent natures. Human history proves that they have good reason to fear us. To this day Air Force pilots have a standing order to shoot down any UFO they come across.

UFO FACTOID
Military officials, such as General Douglas MacArthur, have long held that an interplanetary war with aliens is inevitable, and that the world's

As I write this sentence in the year 2010, my own country, the U.S., is at war once again and thousands of lives have already been lost, with more sure to follow. Most of these conflicts actually serve no purpose, and there is never any real winner. They only create more hatred, pain, destruction, misery, and death. But we fight anyway. We are indeed a violent species.

So we've proven to the aliens—and we continue to prove to them—that we're a potentially dangerous creature.

Pop culture supports this view. In most of the movies we make about aliens, such as *The War of the Worlds*, *The Invasion*, *E.T. The Extraterrestrial*, *District 9*, *Species*, *The Day the Earth Stood Still*, *Mission To Mars*, *Alien*, *Independence Day*, *Night Skies*, *Men In Black*, *Predator*, *Signs*, and *It Came From Outer Space*, we portray them as hideous, vile enemies who must be destroyed.

In reality, many governments do seem anxious to "pick a fight" with aliens. In 1956, for example, the U.S. Navy sent out the following order to all pilots: "If you encounter a UFO that appears hostile you must try to shoot it down." (As we'll be discussing shortly, military documents show that we have fired guns, cannons, and missiles at aliens hundreds, if not thousands, of times.)

Actually, the U.S. government has been treating aliens as if they're our sworn enemy from the very beginning. Here's what the famed American military leader General Douglas MacArthur said in 1955, over half a century ago: One day soon the governments of the world will need to form a single coalition, for the next major world war will be an interplanetary war, one fought between humans and beings from other worlds.

So now ask yourself: if you were from a distant star system, would you want to land out in the open on planet Earth?

CHAPTER 2 AT A GLANCE

- Like the objects they travel in, aliens come in almost every shape, size, color, and description.

- Though there are currently at least sixty reported types of aliens, an ET is most likely to belong to one of eight basic, but different, varieties or species.
- We know aliens are real because millions of people (called "experiencers") around the world have seen them, touched them, smelled them, and heard them.
- Aliens seem to average about four feet in height—at least the ones we know about.
- Aliens are secretive, strange, powerful, highly evolved creatures who may be 1 billion years older than humans.
- Aliens appear to operate under the leadership of an all-powerful female or goddess-like figure called "the One."
- Aliens may have helped us evolve from prehistoric primates to modern human beings.
- Aliens may be genetically related to us in some way.
- Aliens may be here, in part, on a mission to help stop us from destroying ourselves.
- Aliens usually communicate telepathically, or even through their eyes. But they can also speak verbally.
- Aliens usually have no aroma. But they can smell like wet paper, cinnamon, cheese, or rotting meat.
- Aliens are known to have very soft rubbery skin, something like that of a whale or dolphin.
- Aliens have few if any teeth, seem to eat very little, and may live off of a strange, green vegetarian liquid.
- Aliens have astonishing powers that defy what we know about the laws of physics.
- Aliens are abducting humans and performing medical tests on them.
- Aliens are afraid to land out in the open on our planet because they see us an aggressive and violent species.

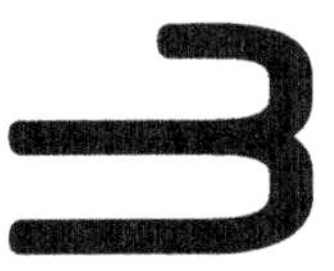

THE PAST, PRESENT, & FUTURE OF UFOS

"A circle of fire coming in the sky, noiseless, one rod [16.5 feet] long with its body and one rod wide. After some days these things became more numerous, shining more than the brightness of the sun." — Egyptian hieroglyphs dating from the reign of Thutmose III (1504-1450 BCE)

ALIENS HAVE BEEN ON EARTH FROM THE BEGINNING

UFOs and aliens are not a modern phenomenon. They've actually been visiting us for hundreds, perhaps even thousands, of years. In fact, they may have been here on Earth from the very beginning.

For humanity's very earliest artistic records show that long ago—long before even the invention of the wheel—people were encountering extraterrestrial beings.

UFOS & ALIENS IN PREHISTORIC TIMES

Some 15,000 years ago, for instance, a Paleolithic French man or woman walked into a cave one day and began drawing on one of its walls. But the picture he or she made wasn't of a bison or a horse, or one of his family members, as you might expect. It was a drawing of a formation of disk-shaped objects flying through the air.

Around 7,000 years ago an unknown artist in Pakistan painted a helmeted creature getting out of a hovering saucer-shaped craft. And some 3,000 years ago, on the Peruvian plains of Nazca, Native Americans created huge pictures in the soil that are indecipherable on the ground and which can only be seen and recognized from the air.

How do mainstream scientists explain these things?

So far they haven't been able to: all of these images were made long before humans learned to fly. For those who accept the reality of UFOs, however, the answer is obvious.

UFOS & ALIENS IN ANCIENT MYTHS

> UFO FACTOID
> Prehistoric people painted what appear to be helmeted space beings on the walls of caves thousands of years ago. African rock paintings of alien-like creatures, for example, date back some 8,000 years, while in Italy a cave painting was found of helmeted humanoids that's at least 12,000 years old.

Although today we refer to the occupants of UFOs as "extraterrestrials," "EBEs," or "aliens," early people saw them as supernatural creatures or even deities. Thus they gave them such names as "goddesses," "gods," "angels," "demons," "fairies," and "elves."

Stories of divine beings who "came down from the stars" are found in the ancient records of almost every land, from Japan and North America to Israel and Australia, from Ireland and Iraq to Hawaii and Germany. Let's look now at a few examples of ancient records of close encounters between people and aliens, records we call "myths."

The ancient Egyptians taught that the chief female alien was a goddess named Isis, and that she lived on the planet Sirius. The chief male alien was a god called Osiris, said to dwell in the constellation of Orion. The Egyptians also believed that the sky is a huge ocean and that Isis and Osiris sailed across it in "sky ships." As the quote at the beginning of this chapter shows, writings from 3,500 years ago state that one day, during the reign of King Thutmose III, the Egyptian people saw "fire circles" that were "brighter than the Sun" come flying down out of the sky.

In ancient Greek myth we find the tale of a man named Phaeton, who rode a "chariot of the Sun" into the sky. As he rose higher and higher, he was shot down by a laser-like bolt of light from the Greeks' chief male alien, the god Zeus, the same Supreme Being known to the ancient Romans as Jupiter, to Jews as Yahweh, and to Christians as Jehovah. Today this being is most commonly referred to simply as "God."

From ancient Rome there comes the story of a man called Menippus, who used eagle's wings to fly to the Moon, then to the Sun, then on to Heaven itself. Heaven was the home of the Romans' chief female alien, the goddess Hera, who became angry at Menippus for intruding on her. So she asked another alien, the god Hermes, to capture him. Hermes tore off Menippus' wings, dropped him over the edge of Heaven, and watched as he fell all the way back down to Earth.

UFO FACTOID
The Native American Kayapo of Brazil are just one of the many primitive peoples who created artwork of helmeted, space-suit wearing beings. The Aztecs portrayed their chief god Quetzalcoatl (whom the Mormons believe is Jesus) as riding in an intricately made spacecraft, and early Australian aborigines created portraits of modern looking space travelers.

The ancient Scandinavians taught that the Northern Lights are caused by light reflecting off the armor of a band of female alien warriors who they called the Valkyries. According to early Nordic myths, the Valkyries were goddesses who charged across the night sky on their beautiful horses.

In ancient Mexico, Native American myth tells the story of how a divine rock fell down from the sky, cracked in half, and poured out hundreds of immortal beings. These creatures, whom early Native Americans called "Goddesses and Gods," then went on to populate the entire Earth, creating humans and all other forms of life.

UFO FACTOID
Many alien-like creatures appear in the myths of ancient cultures, including those in the Bible, where we find a strange group of winged beings called the cherubim (Genesis 3:24).

In ancient China there's a legend about "the Land of Flying Carts," where three-eyed alien-like beings drove "winged chariots" that sailed through the air.

East Indian myths speak of "divine messengers," who rode in "heavenly vehicles" and used "star-weapons" against all-powerful "beings of light" who came from the planet Venus, while their sculptures depict all-powerful beings riding in exotic flying machines.

CLOSE ENCOUNTERS IN THE OLD TESTAMENT

The people who lived in the Near East 3,500 years ago also had many encounters with UFOs and aliens. We know this because they wrote about their experiences in the Bible. In fact, the Bible has some of what could be interpreted as the most detailed descriptions of close encounters, unknown spacecraft, and bizarre extraterrestrials, that have ever been recorded.

DDs and NLs, for instance, were commonly sighted by people living in Old Testament times. Of course, they didn't refer to them as DDs and NLs. Instead, using the language of their day, they called them "pillars of clouds," "pillars of fires" (Exodus 13:21-22), and "doves" (Isaiah 60:8), among many other names.

CE1s were common. But again, they didn't refer to them as UFOs. Instead the ancient Hebrews described them as "chariots of fire" (2 Kings 6:17), "smoking furnaces," and "burning lamps" (Genesis 15:17).

The people of the Bible also reported CE2s, CE3s, and CE4s. Many of these were experienced by Moses, a man said to have lived around 1600 BCE.

In one episode Moses and his followers have a CE2 in which a UFO, or "pillar of fire," leads them across the desert at night (Exodus 14:19-29). In another story Moses has a CE3 in which he encounters a heavenly spacecraft with an alien who speaks to him (Exodus 33:9-10).

UFO FACTOID
A Medieval painting of Moses and the tablets portrays a saucer-shaped craft flying through the air in the background.

According to Jewish mythology, in the year 896 BCE, Elijah had a CE4, which the Bible describes this way:

> And it came to pass, as they still went on, and talked, that, behold, there appeared a chariot of fire, and horses of fire, and parted them both asunder; and Elijah went up by a whirlwind into heaven (2 Kings 2:11).

Like the crew of the Starship *Enterprise*, Elijah seems to have been transported, or "beamed up," within a stream of light ("whirlwind") into a bright hovering spacecraft ("chariot of fire").

The same experience happened to a man named Enoch. The Bible states quite clearly that Enoch never died, but instead was taken up, or "translated," into space. And after this he couldn't be found, for "God took him" (Genesis 5:24; Hebrews 11:5).

In the year 519 BCE, Zechariah had a CE1. What follows are his actual words from the King James Bible:

> Then I turned, and lifted up mine eyes, and looked, and behold a flying roll. And [an angel] . . . said unto me, What seest thou? And I answered, I see a flying roll; the length thereof is twenty cubits, and the breadth thereof ten cubits (Zechariah 5:1-2).

UFO FACTOID
A Sumerian clay tablet, known as the Berlin Seal, was made over 5,000 years ago, yet it reveals detailed knowledge of outer space. According to conventional scientists, however, true astronomy didn't begin until a mere 500 years ago, with Galileo, the Father of Modern Science.

From this report it's clear that Zechariah saw what appears to be a large hotdog bun-shaped spacecraft in the sky, which he estimated to be about thirty feet long and fifteen feet wide. Except for the language, Zechariah's description is identical in every way to those given by many modern day UFO witnesses, right down to the dimensions of the craft.

Undoubtedly, the most famous biblical eyewitness account of a UFO comes from Ezekiel, who, in the year 595 BCE, experienced a major CE3. Here's how he describes his encounter in the Bible:

> And I looked, and, behold, a whirlwind came out of the north, a great cloud, and a fire infolding itself, and a brightness was about it, and out of the midst thereof as the color of amber, out of the midst of the fire. Also out of the midst thereof came the likeness of four living creatures . . .
>
> As for the likeness of the living creatures, their appearance was like burning coals of fire, and like the appearance of lamps: it went up and down among the living creatures; and the fire was bright, and out of the fire went forth lightning. And the living creatures ran and returned as the appearance of a flash of lightning.
>
> Now as I behold the living creatures, behold one wheel upon the earth by the living creatures, with his four faces . . . and their work was as it were a wheel in the middle of a wheel . . .
>
> As for their rings, they were so high that they were dreadful; and their rings were full of eyes round about them four. And when the living creatures went, the wheels went with them

> and when the living creatures were lifted up from the earth, the wheels were lifted up. . . . (Ezekiel 1:4-19)

From Ezekiel's description it's clear that he saw a brilliant flying circular object come down from the sky and land on the ground. Four aliens then got out of the spacecraft—"light beings" who were dressed in shiny spacesuits and whose skin glowed with a strange light. They also carried some kind of portable laser, and were able to move about very quickly—much faster than a human. Ezekiel describes the UFO as two spinning rings with lights around their edges, that finally shot up into the sky.

> UFO FACTOID
> The famous Dogu statues depict astronaut-like beings wearing space suits. Incredibly, they were made some 12,000 years ago by a Neolithic Asian people called the Jomon.

EXTRATERRESTRIALS & SPACE WEAPONS IN THE NEW TESTAMENT

The New Testament also records numerous human encounters with UFOS, alien-like creatures, space weapons, and mysterious lights taking people up into the air.

It seems very likely that around 2,000 years ago, the Three Wise Men followed a NL, or unidentified Nocturnal Light, to the house of the baby Jesus (Matthew 2:7-10). The shepherds nearby had a CE4 when an alien "came upon them" while they were in the fields. The brightness of the bizarre being absolutely terrified them (Luke 2:8-9).

Some 1,950 years ago it would appear that Saint Paul encountered a DD on the road to Damascus (Acts 22:6), and later he seems to have had a CE4 in which he was taken "up into the third heaven" (2 Corinthians 12:2-4).

> UFO FACTOID
> "The Annunciation with Saint Emidius," a painting created in 1486 by the Italian Renaissance painter Carlo Crivelli, clearly shows a UFO-like object in the sky shooting down a beam of light. What is it?

While on a mountain top it looks like Jesus and three of the Disciples experienced a CE4, in which an alien being came down in a mysterious aircraft and talked to them (Mark 9:2-8). The canonical Gospels state that after Jesus was baptized, a strange ghost-like creature came down from space and took on the appearance of a dove (Luke 3:22).

In another Gospel report we find a CE3, in which a light being descends from the sky to Jesus' tomb. The alien's face glowed "like lightning" and his clothing was as "white as snow" (Matthew 28:2-3). Jesus himself then became as "bright as lightning," after which he was "carried up into heaven" (Luke 24:51). This could be seen as a CE4 in which Jesus was abducted by a spacecraft as two luminous aliens stood by (Acts 1:9-11).

At times Jesus himself seems to be portrayed as an all-powerful extraterrestrial being, such as when he's referred to as "the Sun of Righteousness" (Malachi 4:2), a common title taken from numerous popular pre-Christian gods, like the Pagan deity Mithras, born of the Virgin-Mother Anahita on December 25.

> UFO FACTOID
> The Medieval Visoki Decani fresco called "The Crucifixion," shows rocket-like aircraft in the sky on either side of Jesus' cross.

Indeed, the Bible is rife with allusions to Jesus as a blinding, shining, omnipotent divinity from the Heavens (see for example, Isaiah 9:2, 42:6; Matthew 4:16; Luke 1:79; John 1:4, 9; 8:12; 12:35, 46; 2 Corinthians 4:6; Ephesians 5:14; 2 John 2:8; Revelation 21:23), in every way identical to the Sun-Gods of early Paganism—beings who may have been high level alien leaders of some kind.

The book of Revelation tells us that "in the last days" all of the good people will be taken up to heaven "in a cloud" (Revelation 11:12), and that all of the wicked people will be devoured by a fire that will come down from out of heaven (Revelation 20:9).

In another instance of alien teleportation, Christian legends say that the Virgin Mary never died, but instead was "lifted bodily up into heaven" in a strange and brilliant light.

"MANIKINS," "GREEN CHILDREN," & SILVER DISKS: UFOS & ALIENS FROM THE MIDDLE AGES TO THE 16th CENTURY

During the Middle Ages (from the year 500 to the year 1500) UFOs continued to be seen all over the world, as the art from this period shows.

Medieval monks living in a monastery in Yugoslavia created a fresco showing a spacecraft speeding through the sky with a man sitting in it. Around the same time Russian church artists painted two jellyfish-like UFOs with occupants soaring above a crucifix of Jesus.

But this is just artwork. There were also real sightings, and even close encounters with aliens, recorded from this period.

One day, during the 4th Century, Saint Anthony of Alexandria was walking through a canyon when he came upon a small but queer space-creature that he described as "a manikin with a hooded snout, horned forehead, and extremities like goat's feet."

In the 9th Century French soldiers were astounded to see ghost-like creatures they called "tyrants of the air" flying formations across the sky in "wonderfully constructed aerial ships."

Around the year 1211, in Cloera, Ireland, a group of people were attending church when they were interrupted by a loud bang on the roof. Rushing outside, the parishioners looked up, only to see a huge "ship" floating in the sky overhead. It had accidentally dropped some kind of "anchor" on top of the church.

UFO FACTOID
Another painting from the 15th Century, "The Madonna with Saint Giovannio," by the superb Italian artist Domenico Ghirlandaio (Michelangelo's teacher), features a glowing disk-shaped object in the sky.

In Medieval England we have reports of two "green" children, who wore peculiar clothing, and who came from a "twilight land" where there was no Sun. In 1290, again in England, a group of terrified monks reported seeing an enormous silver disk that flew silently over their monastery.

Paintings from the 1400s show saucer-shaped objects traveling in the sky, unusual craft which the witnesses called "sky galleons." The 16th-Century seer Nostradamus, said to have seen both UFOs and aliens, predicted that in 1999 "a great king of terror will descend from the skies."

That year has long since passed. Who or what was he referring to? Many sightings of UFOs and aliens were reported that year. Was he speaking of one of these specifically?

SCOTTISH ALIENS & "AIRSHIPS" OVER AMERICA: UFOS FROM THE 18th TO THE 19th CENTURIES

In the 1700s many famous scientists saw unidentifiable things flying through the sky.

English astronomer Edmund Halley watched a strange object for two hours on the night of March 16, 1716. It was so bright he could read a book by its light.

UFO FACTOID
Sightings of UFOs have continued nonstop from prehistoric times to the present. Eyewitnesses living in the 1700s and 1800s called them "airships" and "sky galleons."

In 1744 twenty-six people in Scotland stood for two hours watching a large group of weird looking beings soaring mysteriously over a mountain. Around this same time the great philosopher Emanuel Swedenborg was abducted by some sort of alien creatures, an experience that he later wrote about in great detail.

On August 9, 1762, Swiss astronomers Monsieur de Rostan and M. Croste saw an object pass in front of the Sun. And on June 17, 1777, French astronomer Charles Messier saw "dark spots" in the sky that he described as large, swift "ships" shaped like bells.

In the 1800s UFOs were spotted over England, Malta, Switzerland, Germany, and Portugal. Brightly lit objects passing in front of the Moon were also reported at this time. In 1824 the astronomer Gruythuisen observed lights actually moving around *on* the Moon.

In 1878 a man in Texas saw a speeding object in the sky that he described as a "large saucer." In 1893 hundreds of people, from Virginia to South Carolina, watched as a huge UFO flew through the night sky.

In 1896 one of the most intriguing and lengthy UFO cases in history began. All over California hundreds of people started seeing cigar-shaped "airships" which darted about "at incredible speeds," and hovered motionless in midair, even in high winds. Witnesses also saw NLs accompanying the craft, most shaped like balls, eggs, and "V's."

UFO FACTOID
In 1903 the Wright brothers altered history with the first successful manned flight on the beaches at Kitty Hawk, North Carolina. But UFOs were seen long before this.

Many people reported that the airships were covered with flashing lights, and that human-like beings inside used searchlights to examine the ground. These eerie UFOs traveled across America and then into Europe where they were spotted well into the 1940s, the time period which marks the start of the modern Age of UFOs.

UNIDENTIFIED FLYING OBJECTS EXISTED BEFORE THE INVENTION OF FLIGHT

What is so amazing about these early sightings of UFOs is that they all occurred before we humans knew how to build and fly self-propelled aircraft.

It wasn't until December 17, 1903, that Orville and Wilbur Wright made humanity's first successful flight in a motor-powered airplane. The Wright brothers' journey through the air lasted a mere fifty-nine seconds, and their charming but flimsy plane traveled only 852 feet. That's how primitive our aeronautical technology was in the early 1900s. And it would be many more years before we'd learn how to build such things as helicopters, jets, and space vehicles.

These facts beg several questions:

UFO FACTOID

On a winter night in 1942 a large UFO appeared menacingly over Los Angeles, California. It was repeatedly pounded by military artillery, but the shells simply bounced off the object. Eventually it slowly moved away and disappeared into the night.

- Why did prehistoric people, living thousands of years before even the discovery of writing, create images that look identical to modern UFOs?

- Who were the all-powerful radiant beings that the ancient Egyptians, Greeks, and Jews saw descend from the sky, then float back up again?

- If they weren't UFOs, what were the swiftly darting, brightly glowing hovering objects that Europeans saw in the Middle Ages, and which Americans saw later between the 17th and 19th Centuries?

THE BEGINNING OF THE MODERN UFO ERA

Unearthly cigar- and disk-shaped flying objects continued to be seen all over the world throughout the 1920s. And in the 1930s Europe was finally overrun with UFOs.

Casebook Study 7: Scandinavians, for instance, began reporting what they called "ghost rockets" or "mystery flyers," odd unmarked aircraft that performed close-to-the ground maneuvers—in total silence. Not only this, these objects flew easily in severe and hazardous weather, such as fog, rain, and snow, conditions that no human-made aircraft could even take off in, let alone fly in.

Casebook Study 8: On February 25, 1942, during World War II, unearthly aircraft began to be seen over Los Angeles, California. Thinking the objects were Japanese fighter planes, the military sounded air raid sirens around the city and began firing at them with artillery guns and cannon. But to the soldiers' amazement the antiaircraft shells bounced right off the sides of the objects.

Thousands of people watched one large object, bathed in searchlights, hover in the air as bombs landed right on it and explode. Incredibly no damage was done. Then, when the military stopped firing on the stationary craft, it drifted slowly away as if nothing had happened, disappearing over the horizon as if it was on a leisurely tour of the California countryside. Military officials at the time said that the objects were "in fact not earthly and, according to secret intelligence sources, they are in all probability of interplanetary origin."

Casebook Study 9: In 1945, in the middle of World War II, strange glowing balls of lights, nicknamed "foo fighters," were spotted pacing, overflying, and darting around the planes of American, German, Japanese, and British fighter pilots.

Though their behavior understandably alarmed many pilots, the weird objects caused no harm. Indeed, the foo fighters (from the French word *feu*, meaning "fire") showed up again in later conflicts, such as the Korean War and the Vietnam War, seemingly to keep pilots company. Still no one really knew what any of these unexplainable flying objects were. There wasn't even an agreed-upon phrase to describe them yet.

UFO FACTOID

Foo fighters, bizarre balls of colored light, often accompanied American, British, German, and Japanese military planes during World War II. Performing aerial maneuvers at speeds that seemed to defy the laws of physics, all efforts to identify the objects failed, and to this day no one knows what they were.

But the answer—and the phrase—finally came on June 24, 1947.

Casebook Study 10: On this day civilian pilot Kenneth Arnold was flying over Mount Rainier, Washington, when he saw nine shiny crescent-shaped objects flying in **echelon** formation, silhouetted against the Cascade Mountains.

They were traveling at an incredible 1,200 mph, and they looked, as Arnold later said in a newspaper interview, like "saucers skipping over water." A journalist thought up the term "flying saucer" to describe Arnold's sighting, and we've been using it ever since.

A few years later the USAF came up with the more scientific term "unidentified flying object," or "UFO." This is how and when the modern era of UFOs began.

FROM ORANGE BALLS TO FLYING WINGS: UFOS TODAY

From 1947 to the present day not only have extraterrestrial craft continued to visit Earth, but sightings of UFOs have increased ten times since World War II (1939-1945). Why did the aliens find us more interesting than ever after the War?

UFO FACTOID
Kenneth Arnold's sighting of a group of fast-moving unknown objects over Mt. Rainier, Washington, in 1947, launched what's known as the "Modern UFO Era."

Possibly because we dropped the first atomic bomb in our planet's history on August 6, 1945, over Hiroshima, Japan, eventually killing 242, 437 people. This is certainly something that got not only the world's attention, but most probably the attention of beings living on other worlds as well, since the flash of the explosion could be seen from space.

The aliens' increased interest in us became all too apparent when, over several weeks in July 1952, a major UFO **flap** occurred in the political heart of the most technologically advanced nation on Earth: Washington D.C., the capital of the U.S.

Casebook Study 11: It was at this time that large formations of what the newspapers humorously called "aerial whatzits" were seen over the city. There were so many UFOs in the sky, hovering, diving, and making ninety-degree turns, that at least 2,000 known sighting reports were logged by concerned citizens. Some tried to brush off the event, casually referring to it as the "Washington merry-go-round." But in reality the situation was not fun. And it was certainly no laughing matter.

Over a period of many weeks, thousands of people saw hundreds of unconventional shiny disks and glowing objects flying directly over the White House. Not only were the UFOs photographed, but they were also simultaneously picked up on radar traveling at an amazing 7,000 mph.

UFO FACTOID
After squadrons of UFOs were seen by thousands over America's capital city in the summer of 1952, the *Washington Post* ran this headline: "'SAUCER' OUTRAN JET, PILOT REVEALS."

Harry Barnes, Senior Air Traffic Controller for the **CAA**, made the following statement: "For six hours . . . there were at least ten unidentifiable objects moving above Washington. They were not ordinary aircraft." Albert M. Chop, chief U.S. Air Force press officer at the time, pronounced the bizarre objects "real" and "interplanetary" in origin.

As might be expected, because this UFO wave occurred over the nation's capital, it caused a major scandal. People were afraid that if the U.S. military couldn't protect its own governmental headquarters from "alien invaders," then it couldn't defend the rest of the country either.

In answer to these fears high level government officials said that the objects weren't UFOs at all: they were "masses of electrically charged particles of air."

The American public wouldn't accept this obviously ridiculous answer, and pressure finally forced the government to hold a special session at the Pentagon to investigate the sightings. But the outcome was not what anyone expected: secret Pentagon meetings (held January 14-17, 1953) concluded that "all UFOs could be explained and identified." (In Chapter Six we'll learn why our governments continue to hide the reality of UFOs from the public.)

UFO FACTOID
True UFOs are obviously made and operated by intelligent beings. The question is, are they from Earth or somewhere else? Overwhelming evidence suggests the latter.

In the 1960s and 1970s UFO sightings in the U.S. seemed to decrease. But not because UFOs weren't visiting us. They were here alright. America was preoccupied with a number of social and political problems, such as the assassinations of President John F. Kennedy, Martin Luther King, and Robert Kennedy, Watergate, the Civil Rights Movement, the Women's Movement, and the Vietnam War.

In the 1980s things began to pick up again, this time stronger than ever before.

Casebook Study 12: For example, on May 5, 1984, American satellites—stationed in orbit some 22,000 miles above the Earth's surface—detected an unidentified flying object enter Earth's atmosphere from outer space.

Scientists were completely baffled by the bizarre object, which they later code-named "Fast Walker." Why the puzzlement? Because instead of continuing to fly toward the surface of our planet, as one would expect it to, the object abruptly reversed direction and traveled back out into space!

There is no natural inanimate object that has the ability to fly into Earth's atmosphere, then turn around and fly back out again. This is impossible. Only a self-propelled vehicle made and operated by extremely intelligent beings could do this. And no country on Earth had any spacecraft capable of performing this maneuver in high orbit in 1984.

> UFO FACTOID
> Numerous black, triangular UFOs were seen over Belgium in 1989. Skeptics explain them away as "secret military aircraft." But flying triangles were seen in our skies long before the modern era.

So what was Fast Walker? Even the most skeptical scientists still can't explain what it was. But we can: it was a genuine UFO.

RECENT UFO SIGHTINGS

Here are some more exciting examples of recent UFO sightings, taken from authentic reports.

Casebook Study 13: In the late 1980s hundreds of people in Gulf Breeze, Florida, began seeing UFOs hovering and streaking across the sky. Both DDs and NLs were spotted. Some looked like a child's top, others like silver saucers.

One night eyewitnesses videoed a spinning UFO flashing green, red, and white lights, drop a smaller UFO down to the ground. On another night onlookers clearly observed several UFOs zoom into view, dart around the night sky, then simply "blink off."

Casebook Study 14: In January 1992 people in Arkansas were astonished to see a giant "flying wing" gently drifting overhead a mere 1,000 feet off the ground. It had a number of bright lights on it and was completely silent. Eyewitnesses say that it was "bigger than an aircraft carrier."

Casebook Study 15: In April 1992 in Washington, D.C. a number of people watched eleven UFOs dart around above the Washington Monument for twenty minutes. The silvery craft then flew high into the sky, jostled themselves into formation, and shot off into the distance.

UFO FACTOID
Who's sitting in the driver's seat of the thousands of UFOs that are spotted around the world each year? We don't have the kind of technology needed to design, build, and operate these types of aircraft.

Casebook Study 16: In July 1992 scientists in Wiltshire, England, spotted four large orange balls of light moving silently across the night sky. The objects moved in a "jerky" manner and were each about thirty-five feet across.

Casebook Study 17: In November 1996 a large UFO that looked like a miniature spinning galaxy was seen all over South Korea. The mysterious craft glowed with a strange light and hovered in midair before blasting into the sky and disappearing.

Casebook Study 18: In March of 1997 dozens of UFOs were seen around New England, hovering over expressways, playing "catch-me-if-you-can" with motorists.

Casebook Study 19: Also in March 1997, this time in Phoenix, Arizona, people stood in utter disbelief as a huge black V-shaped UFO, four city blocks wide, drifted silently over their suburban homes. It was so close to the ground that they could see the minute details of its surface.

Called "the biggest UFO sighting in history," the giant triangular object traveled over 200 miles across the Arizona skies, frightening the public and baffling military experts.

Eyewitnesses, which included police officers and air traffic controllers, testified that the slow-moving object sported red and white lights and was the size of at least three football fields. The mysterious craft was visible for two hours, and hundreds of people took photos and video of it.

UFO FACTOID
The Phoenix UFO incident in 1997 stunned the nation and made believers of many skeptics.

Skeptics dismissed the sighting as "military flares," an obvious impossibility since flares burn out and fall to the ground in irregular patterns in just a few minutes (see the next casebook study).

Casebook Study 20: This same weird craft was seen again in June 1997 drifting over Las Vegas, Nevada. Bewildered eyewitnesses say that it made no noise and had six bright lights on its underside. When three fighter jets from Luke Air Force Base were scrambled to intercept the object, it launched straight up in the air at blinding speed and disappeared from view.

Casebook Study 21: On March 5, 2004, Mexican Air Force pilots were flying a Merlin C26A bimotor airplane at 11,500 feet over the state of Campeche, Mexico. On a routine anti-narcotics run, they were in search of drug smugglers, a major cause of crime in the area.

At about 5:00 PM they suddenly found themselves surrounded by eleven bright UFOs. Strangely, of the eleven craft, only three showed up on radar.

The pilots, highly trained in aircraft identification, had no idea what they were, and described them as looking like "sharp points of light," and at other times like "large headlights."

UFO FACTOID
Mexico has one of the highest densities of UFO sightings in the world, many seen by millions of people at the same time.

The UFOs seemed to be intentionally cloaking themselves in blurry light, and all moved rapidly around the military jets, easily evading capture when the pilots tried to pursue them.

As always, skeptics laughed off the sighting. But the unknown objects were videotaped by the frightened pilots, who later stated that whatever they were, they were very real and appeared menacing, even causing them to go to "red alert."

Mexico's Department of Defense turned over a copy of the amazing tape to the media and it was aired around the world on international TV to millions of viewers.

These are just a few of the hundreds of thousands of UFO sightings that have been reported in the past few decades alone.

Here's an example of a very recent, and disturbing, **multiple sighting**.

Casebook Study 22: On November 7, 2006, at Chicago's O'Hare Airport, at least a dozen United Airlines employees, including pilots and mechanics, looked up and saw a dark gray, metallic, saucer-shaped object hovering motionless at about 1,900 feet above the terminal complex. After a few minutes the frisbee-like craft shot straight up into the sky, punching a perfect round hole in the clouds.

The shaken eyewitnesses filed reports with United, but the airline refused to investigate the sighting. Neither the **TSA**, the Chicago Department of Aviation, or the **FAA** showed any interest either, the latter stating that it was no doubt just a "weather phenomenon."

THE FUTURE OF UFOS

We've seen that UFOs have been visiting Earth for as long as humans have been here, and that they're still very much interested in us. So what does the future hold for UFOs?

One thing is certain: UFOs will continue to be seen all over the world by more and more people. Why? Because today these strange craft, and their even stranger occupants, are coming to our planet in greater numbers than ever before. In Great Britain, for example, while over 11,000 unidentified flying objects have been sighted in the past thirty years alone (that's 366 a year, or about one a day), even more are being reported by Britons today.

Also our scientific knowledge is growing, which lets us build better telescopes, cameras, and radars, and at less cost, making seeing and recording UFOs easier and more accessible.

The future of the field of ufology itself is also very exciting, since it's growing and changing on a daily basis. Authentic new UFO documents, new eyewitnesses, new UFO reports, new UFO photos and video are constantly coming to light. All of these things are helping us to get a clearer picture of what UFOs are, why they're here, and who's piloting them.

UFO FACTOID
Around the world UFO sightings are increasing in number each year.

However, as more and more people are seeing UFOs, and believing in them, skeptics are becoming more and more determined to debunk them. Actually, skeptics have a very different future in mind. They'd like to see the entire idea of UFOs and aliens thrown out and forgotten.

But the fact is that "seeing is believing." This is why skeptics have already lost their fight to convince the world that UFOs don't exist. Thus, interest in UFOs is growing worldwide and the future of ufology is very promising.

Each day many new believers are added to the millions who are already convinced that we are not alone. Subscriptions to UFO magazines, memberships in UFO organizations, and attendance at UFO conferences are increasing every year. And there's no sign of this interest slowing down.

Tourists travel from all over the world to visit UFO "hot spots," and the Internet is filling up with thousands of sites dedicated to the topic of aliens.

The extent to which the public accepts the reality of UFOs and aliens is reflected in the fact that they've even entered the mainstream, appearing all over the world in TV commercials and magazine ads.

Entire TV shows, like *Hangar 1*, *Chasing UFOs*, *Ancient Aliens*, *UFO Hunters*, *UFOs: The Untold Stories*, *The X-Files*, *Stargate*, *Dr. Who*, *Battlestar Galactica*, *Babylon 5*, *Roswell*, and Gene Roddenberry's *Earth: The Final Conflict*, are (or were) dedicated to the theme of UFOs and aliens, and more and more films, like *Dark Skies, The Fourth Kind, Night Skies*, *Independence Day*, *Men in Black*, *Contact*, *Terminator*, *Predator*, *The Abyss*, *Sphere*, *The Forgotten*, and of course *Alien*, have been, and continue to be, made about this topic.

UFO FACTOID
UFOs have always been with us, and will continue to be with us long into the future, no doubt, in fact, for as long as humanity exists.

There's now a TV channel devoted, in part, to the belief in alien reality, the Syfy Channel, while many TV programs, such as *The Unexplained Files*, *NASA's Unexplained Files*, *Fact or Faked: Paranormal Files*, *Monster Quest*, *Haunted Highway*, *History's Mysteries*, and *Mystery Quest*, regularly include pieces devoted to the scientific research of UFO-related events and phenomenon.

Even wholly scientifically-oriented TV networks, like the Science Channel, the Discovery Channel, the Military Channel, the National Geographic Channel, and the History Channel, now regularly air programs about UFOs and aliens, often without any skeptical commentary at all. Ufology has indeed gone mainstream.

Could it be that ETs are secretly behind all of this? Are they slowly preparing humanity for the day of "first contact" between our species and theirs? Are they trying to get us used to them through TV and movies so that we won't be so shocked when they actually do land on the White House lawn?

Many people think so and, as we're about to see, there is some evidence that this idea—as unlikely as it sounds—might just be true.

CHAPTER 3 AT A GLANCE

- Aliens and UFOs have been visiting Earth for thousands, possibly millions, of years.
- At least 15,000 years ago prehistoric people were already drawing pictures and carving images of aliens and UFOs.

- Ancient people didn't call extraterrestrial beings "aliens." They referred to them as "goddesses," "gods," "angels," and "demons." And they left us many stories about them that we call "myths."
- In both the Old Testament and the New Testament there are dozens of detailed reports of close encounters with UFOs and aliens.
- People in the Middle Ages called aliens "tyrants of the air" and they called UFOs "sky galleons."
- After the invention of the telescope in the Medieval period, astronomers began to see strange lights, not only on the Moon, but also flying around in outer space.
- A UFO was first described as a "saucer" in Texas in 1878.
- In 1896 people all over the world began to report sightings of cigar- or tube-shaped "airships" that darted around in the sky shining strange lights down on the ground.
- Sightings of UFOs increased dramatically right after World War II, probably because we set off the first atomic bomb in history.
- The modern age of UFOs began in 1947 when pilot Kenneth Arnold spotted nine shiny disks skipping through the air over a mountain range in Washington state.
- In 1952 hundreds of UFOs were seen and filmed flying over the White House at speeds of up to 7,000 mph.
- In the U.S. UFO sightings decreased in the 1960s because people were focused on a number of social problems that were occurring at the time.
- Throughout the 1970s, 1980s, and 1990s, UFO reports of all kinds increased, with more and more close-up sightings of large triangular-shaped "flying wings."
- The future of ufology is very exciting because more and more videos, photos, and documents are coming to light, and because better more precise UFO-detecting equipment is being made.

TAKEN: ALIEN ABDUCTION

"The U.S. Air Force assures me that UFOs pose no threat to National Security." – President John F. Kennedy, 1960s

ABDUCTION: THE SECRET ALIEN AGENDA

UFOs and their alien occupants haven't been visiting Earth for thousands of years because they have nothing better to do. In truth, they seem to have a very serious plan in mind, a plan that we call the "**alien agenda**."

According to some ufologists, one reason for the alien agenda is the fact that the Sun near the aliens' home base is burning out. This is making their own planet too cold to inhabit for much longer, so they're looking for a new one to colonize. To test whether it's possible for them to live here on Earth or not, they're taking soil, plant, animal, and human samples.

> UFO FACTOID
> UFOs are visiting Earth with a secret agenda that many believe involves human DNA and the takeover of our planet.

If this theory is accurate, some believe that our time here on Earth may be limited. The aliens would probably prefer to assimilate into human society and live side-by-side with us in peace. But we've proven to them that we're a highly territorial and aggressive species, and that we'll fight an invasion by outsiders. So, as General MacArthur predicted, the next major world war may be between us and an extraterrestrial race.

But we may have already lost the war even before it starts: as part of the alien agenda they've been testing our defense systems for many years to see if we can withstand an attack from them. It must not

UFO FACTOID
Abductees, humans taken by aliens, often report being subjected to a host of terrifying medical tests and experiments that traumatizes them for life. While the victims' memories are partially or completely suppressed by the aliens, these can usually be fully recovered through hypnotic regression, during which abductees are able to recount every detail of their horrific ordeal. Most retain not only emotional scars, but physical scars on their bodies, as well.

have taken them very long to see that their weapons technology is far superior to our own, and that in an all-out battle with us they would easily win.

No need to worry, however. At the moment this is all just speculation.

A more likely reason for the alien agenda seems to be their desire to create a new superior race that's half alien and half human. The process of mixing the two different creatures to form a new one is called "hybridization."

Aliens may have already achieved this goal. For as you'll remember from Chapter Two, one of the eight basic types of aliens is the Hybrid: a creature that's part human, part alien. (It's possible though that what we call the "Hybrid" is really 100 percent alien, and that it just happens to look as if it's part human.)

Whatever the truth is about the Hybrid alien, there can be little doubt that aliens are harvesting, or taking, DNA (genetic material) from humans and mixing it with their own.

How do they get DNA from humans? Not willingly, we can be sure of that! What human would voluntarily endure a painful operation in order to give a sample of her or his DNA to a creature that looks like a giant insect?

So aliens must take us humans by force, against our will. When an alien kidnaps a human it's called an "**alien abduction**," and people who've been taken in this way are called "**abductees**," or "**contactees**."

UFO FACTOID
The famous painting "The Miracle of the Snow," was created by Italian artist Masolino da Panicale in the 1400s. Yet it shows Jesus and Mary riding inside a sphere on top of a saucer-shaped craft. Why?

Aliens have been abducting humans for as long as humans have been on Earth. We've seen that prehistoric people were encountering aliens at least 15,000 years ago, and that people in the Bible, like Elijah, Enoch, and the Virgin Mary were "translated," or taken up into space in what appears to be a beam of light.

UFO FACTOID
According to Christian legend, the Virgin Mary didn't die a natural death. Instead she floated up into the sky on a long shaft of light.

In modern times alien abduction continues to occur, with many abductees saying that they've been abducted more than once. Still others claim that they've been taken by aliens dozens of times, starting when they were children, and continuing over and over again right into adulthood. This type of victim is called a "**multiple abductee**."

Actually, if the reports are at all accurate, alien abduction is more common today than ever before in history. Four million Americans alone claim to have been taken against their will by extraterrestrial creatures, and that number is growing each year.

The increase in alien abduction may also help explain at least some of the millions of men, women, and children around the world who go missing each year. This equates to thousands who disappear each day and who are never heard from again. Where do they go? Who took them, and why?

THE DISAPPEARANCE OF FREDERICK VALENTICH

In some instances the connection between alien abduction and the disappearance of people is all too obvious, as in the famous case of the Australian pilot Frederick Valentich.

Casebook Study 23: On October 21, 1978, as he was flying his Cessna 182 plane over the ocean near King Island, Tasmania, Valentich noticed a large unidentified object zoom past him from out of nowhere. It had a shiny metallic appearance and was long with four bright lights on it.

The craft overflew his plane several more times, as if it were "playing some sort of game," then settled into a stationary hover pattern directly above his cockpit.

Worried, Valentich called a local airport, only to be told that there were "no known aircraft in the vicinity" besides his.

Then, abruptly, the UFO vanished.

Thinking he was in the clear Valentich radioed back to give his position. But as he was doing so the weird object suddenly reappeared. According to the taped conversation, Valentich then said:

> . . . that strange aircraft is hovering on top of me again . . . it's hovering and it's not an aircraft.

This was the pilot's last transmission. Despite an ocean-wide search no sign of Valentich or his plane was ever found.

WHAT HAPPENS WHEN YOU'RE ABDUCTED?

Skeptics, of course, view the idea of alien abduction as just another humorous claim made by "UFO nuts and kooks." Dismissing them as nothing more than "vivid dreams," they actually see abductions as further proof that there's no such thing as aliens or UFOs.

> UFO FACTOID
> UFOs are no laughing matter: many of them may be searching for humans to abduct. Literally thousands of people around the world disappear *each day* without a trace. According to eyewitness testimony, many of these individuals are the victims of alien abduction.

But to the people who've been abducted, not only is there absolutely nothing funny about it, it's an all too real, and often traumatic, life-altering experience.

Furthermore, there's a pattern to the typical abduction that's experienced by people from different countries all over the world; people who've never met or had any contact with one another. How can this universal pattern be explained if it's not a genuine experience?

Most abductions take place at night, when a person is in bed asleep. But not always.

An abduction can happen anytime, anywhere: during the day on a crowded street in the middle of a city, or at night on a remote country road; when one is with a group of people, or when one is all alone. The aliens operate by their own timetable, not by ours. So abductions occur when it's convenient for them, not for us.

Now let's take a look at the typical abduction. I'll narrate. You'll play the part of the abductee.

It's late at night and you're in bed asleep. You wake up suddenly, feeling a tingling "electrical" sensation over your entire body. You see a blinding bright light outside your window.

Next you see a group of small shadowy figures quickly approaching your bed. You panic. But when you try to scream and run away you find that not only can you not utter a sound, you're completely paralyzed. You can't even lift a finger. Then you hear a voice inside your head saying: "Relax, everything's going to be alright."

By this time you might find that you're more fascinated by what's happening than afraid. Still, you can't move.

Then a beam of light comes through the window, or through the ceiling, and you start to lift off the bed and into the air. You're on your back floating upward on the beam of light, with your arms and legs hanging down limply.

Next, you're totally amazed to find that you float right through the closed window or even right through the ceiling of your bedroom—as if you're a ghost and you don't have a physical body.

Once outside you smell the trees and feel the chill of the night air. You look down and see the ground moving away from you. Your house becomes smaller and smaller.

UFO FACTOID
Skeptics believe that alien abduction is nothing more than a "bad dream." There are several problems with this theory. To begin with the subject is wide awake during nearly the entire abduction. Second, details of the ordeal are vividly remembered for decades after. And third, both the emotional and the psychological impact of the experience remain for life. Dreams, however, occur only when one is asleep, they are forgotten immediately (if remembered at all), and they never affect one's entire life—not even the most vivid realistic dreams.

The next thing you know you're inside a hovering spacecraft, lying face up on a cold metal table. You still can't move. You look around and see that you're in an enclosed round room with medical equipment. The room has a sterile feeling with monochromatic gray walls, floor, and ceiling, yet it's well-lit with bright lights.

The shadowy figures reappear next to you. In the light you can now see what they really look like. They're gray with large heads and huge black eyes. Their bodies look fragile. They have very thin arms and legs, four fingers on each hand, a slit for a mouth, no nose and no hair. They move strangely, stiffly, like robotic insects.

You overflow with conflicting feelings. Although they're terrifying in one way, in another way they're very beautiful to you. You're repulsed by them. But you also feel drawn to them. You want to hate them for what they're doing to you. But at the same time you feel a love for them. They give off a feeling of tremendous peace, but they also seem sinister.

> UFO FACTOID
> Though rare, some eyewitnesses have actually reported seeing other people being abducted, as they were beamed up into waiting UFOs.

They stare at you, looking not just into your eyes, but into your very soul. You feel as if they know all of your thoughts. You actually start to feel that they love you, that they love all of humanity.

At this moment a large TV-like monitor appears. You don't want to look at it, but you can't help yourself. You seem powerless not to gaze at it.

As you do, a video begins to flicker on the screen. Not one showing pleasant images, as you were hoping, but one depicting overpopulation, riots, war, people starving and sobbing, nuclear bomb blasts, polluted rivers, filthy air, parched ground, packed freeways, violent storms, tsunamis, floods, hurricanes, fire, smoke, earthquakes, worldwide misery and mindless violence, cities crumbling into dust or disappearing beneath huge ocean waves.

It's all abhorrent and depressing, and you're wondering why this is being shown to you.

Just as this thought crosses your mind, one of the aliens speaks to you; not with its mouth, but telepathically—from its mind directly to your mind. It answers your question, saying: "You and your kind are abusing planet Earth. These scenes illustrate the future of your world—unless you change your ways."

You realize that you have just been shown several possible end-of-the-world scenarios, and that the aliens are here, in part, to teach us how to avoid these disasters by living more spiritually and ecologically balanced lives.

As you lay there on the metal table overwhelmed with this realization, one of the aliens turns to you and tells you telepathically: "Don't be frightened. We're not going to hurt you."

You're wondering what it's talking about, when all of the sudden the aliens begin to examine and probe you with various medical instruments. They take samples of your skin, hair, and blood. Some of these tests are very scary and even painful.

You scream and cry. One of the gray beings reaches out and touches you in the center of your forehead with its long bony finger. The pain disappears and everything goes black as night.

The next thing you're aware of is being lowered back into your bed on the lightbeam. You fall back asleep and wake up the next day, just like you always do.

But there's a problem: this morning isn't just another morning. Something's different. You have a confused sense that something strange occurred in the night. You just can't remember what.

UFO FACTOID
UFOs pacing our planes, jets, and spacecraft may not be accidental: some pilots report being abducted from their vehicles in mid-flight.

Looking in the bathroom mirror you notice that your pajama top is on inside out. You didn't go to bed like that. Then you feel some discomfort on your shin. Peering down you see a small scoop mark, slightly swollen and red. That wasn't there yesterday.

And there, on your left arm, is a triangular mark, and beneath the skin a small bump. You rub your finger over it and are horrified to feel a tiny hard object under the skin. What's that, and where did it come from?

Now you know something happened last night! But what?

AFTER THE ABDUCTION: BLOCKED MEMORY & MISSING TIME

How could you not remember something as bizarre, frightening, and painful as an alien abduction? It seems impossible.

The reason is quite simple: while you were still on their spaceship the aliens somehow blocked or erased your memory of the experience, giving you a form of **amnesia**. This is why most abductees don't consciously remember their abduction.

Instead, usually little bits and pieces of the memory gradually come back to them as time goes by. Some never remember anything. Others have a dim awareness that something happened to them, but they can't remember exactly what it was.

Some abductees go to a hypnotist and get "regressed"; that is, they have a hypnotist take them back in time so that they can relive the details of their abduction. Then and only then do they begin to realize what really happened to them.

Strangely, some abductees do recall nearly every detail of what occurred during their abduction, even without hypnosis. But this is quite rare.

> UFO FACTOID
> Our governments are keenly aware of UFOs (military aircraft have been seen accompanying UFOs, for example), and may be working in conjunction with ETs under an agreement that allows them to abduct humans in exchange for alien technology.

In many cases an abductee will later see, hear, or read something that triggers a memory recall of their abduction. They might see a TV program on UFOs, or read a magazine article on alien abduction. Almost anything related to unidentified flying objects and extraterrestrial beings can start to bring a piece of the submerged memory back to the surface.

Another intriguing aspect of being abducted is the **MTE**, or "missing time experience." Afterward a person usually feels that the entire experience lasted for only a few minutes. It's only later that they realize that a huge block of time has passed that they can't account for. When they check their watches or clocks they actually find that hours, or even days, have sometimes gone by during the time span of their abduction.

We don't know how to explain this unusual sense of timelessness, but it seems to be related to the aliens' ability to distort space and time.

SCOOP-MARKS & IMPLANTS: HOW ALIENS KEEP TRACK OF HUMAN ABDUCTEES

> UFO FACTOID
> How safe are you from being abducted by extraterrestrial beings? Perhaps not as safe as you think: aliens have been known to abduct people from their cars in heavily populated areas in broad daylight.

After their abduction many abductees find that there are marks, such as scratches, sores, burns, holes, and cuts, on various parts of their bodies. The most common of these is a small triangular mark that feels like a scald and may itch for several days.

Another common type of abduction indicator is the scoop-mark: a hollowed out depression in the skin, usually located on the shin of either leg where there is no fat or muscle to penetrate. It's believed that this is a remnant of an alien "punch biopsy" in which an instrument is pushed through to the bone marrow in order to get at the valuable DNA material inside.

Some abductees don't remember anything about their abduction until many years later, when they discover a strange scar somewhere on themselves that they've never seen before. Sometimes the scar alone can jog an abductee's recollection, bringing to the surface long buried memories.

In many instances an abductee finds that under their scar there's a small bump. When the object is surgically removed it usually turns out to be a piece of metal. But this is no ordinary piece of metal. This is an **implant** that the aliens intentionally put under the abductee's skin.

Even though an abductee may remember being implanted, the implant itself is often difficult to find, even when the abductee's body is X-rayed. This is because the aliens usually create their implants from the same organic minerals that a human body is made of. This is why implants often don't show up even when scanned with electromagnetic radiation (that is, an X-ray).

Skeptics say that this proves that the implants are not of extraterrestrial origin, that they were made by humans on Earth. But the truth is that the aliens use terrestrial material on purpose so that the implants will be difficult to discover. Even when they use materials (mainly metals) that are not common on our planet, the tiny devices don't cause an inflammatory response, so the body never tries to reject them. There's also no visible entry incision on the skin. It's clear that the aliens don't want us to find their implants or remove them.

UFO FACTOID
It's not known why aliens seem to prefer abducting emotionally troubled individuals.

This leads to an obvious question: what's the purpose of an alien implant?

For some unknown reason, aliens like to keep track of the humans they abduct. In fact, sometimes aliens seem to select specific people for lifelong observation, especially, it seems, those who have been emotionally damaged in some way. They begin abducting these individuals from childhood onward, often right into old age.

We don't know why the aliens do this, but we know how they do it: they "tag" the person with a transponder-like implant, in the same way that we tag wild animals like fish, birds, deer, bears, and whales.

This tag is very similar to the bar codes we use on our commercial products: it seems to be encoded with both a computerized number and some kind of a medical evaluation detector, that records information that's important to the aliens. This would probably include such data as a person's body temperature, pulse rate, and blood pressure.

> UFO FACTOID
> Some aliens seem to think of us the way we think of bugs and rodents. Others view us a bit more respectfully, perhaps the way we see elk, wolves, or even chimpanzees. Apparently no alien species sees us as equals or superiors.

The implant also seems to be used by the aliens as a homing mechanism, to trace the whereabouts of their abductees. When they want to find a particular abductee, they simply follow the tracking signal of the implant, which has been placed under the person's skin, or inside their head (inserted up through the nose).

PSYCHIC ABILITIES, PARANOIA, & OTHER AFTER-EFFECTS OF ABDUCTION

A person who's been abducted can expect to experience many different feelings and effects afterward. These can range from the minor to the major, from the trivial to the serious. But whatever these effects are, no one who's been abducted goes away from it unchanged.

Actually, most abductees find that they've been transformed in many deep and mysterious ways. This is yet another piece of evidence showing that alien abduction is genuine: dreams, fantasies, and illusions don't have the power to completely transform us. Only real experiences do.

For instance, many abductees find that they have psychic abilities that they didn't have before, such as being able to see into people's souls, read their minds, or feel their emotions. Other abductees wake up to find that they have the power to heal, or that they've become more spiritual, compassionate, or loving.

Still others find they have a new appreciation for all living things and many of these individuals become vegetarians. Others feel that they have a "mission" in life, one that involves helping others, working with children, senior citizens, or animals, or saving our planet.

Many abductees are so changed by their experience that they go on to become environmentalists, politicians, or religious teachers, occupations where they feel than can affect societal changes.

But not all of the after-effects of alien abduction are positive. Many abductees become quite distrustful and suspicious of the government, or of organizations in general. They sometimes lose all trust in people, see conspiracies in everything, and live in constant fear of being abducted again.

Other abductees are left with a life-long fear of doctors, needles, blood, hospitals, or bright lights. Still others become ill with mental and physical diseases that seem to be directly related to the medical tests that the aliens performed on them. These health problems may include: depression, diarrhea, skin disorders, temporary paralysis, stiffness, numbness, dizziness, headaches, and nausea. Worse, some may turn out to be life-threatening, requiring serious long-term medical treatment.

> UFO FACTOID
> After being abducted the experience stays with a person for life, and some people discover that they have newly found special powers.

Some abductees find the reality of their abduction so horrifying that they bury it deep inside their minds in what the German psychiatrist Sigmund Freud called a "**screen memory**."

In a screen memory the abductee's mind soothes away the pain and fear of the experience by turning it into one that's harmless. The aliens become human "nurses," their spaceship becomes a "hospital," and the alien medical procedure becomes an innocent "visit to the doctor." In this way the screen memory makes it easier for abductees to deal with the intensity of their abductions.

THE ALIEN ABDUCTION OF BETTY & BARNEY HILL

Now that we've studied the basics of alien abduction, let's look at a few of the more spectacular modern cases.

Casebook Study 24: One of the earliest and most astonishing of these occurred on September 19, 1961, in northern New Hampshire. On this night Betty and Barney Hill were driving on Route 3 through the White Mountains, returning from a trip.

UFO FACTOID
Barney and Betty Hill's alien abduction occurred long before the release of films like *Close Encounters of the Third Kind*. This means that the aliens they described under hypnosis weren't based on Hollywood images, as skeptics claim.

All of the sudden, they stared in amazement as a large saucer-shaped object appeared in front of them. The strange craft had two rows of windows, rotating red lights, and wobbled as it came down out of the sky.

Thinking it was a satellite, the Hills continued to drive. But when it finally landed near the road, Barney's curiosity got the better of him.

Quickly pulling over he stopped the car, jumped out, and ran toward it to get a closer look. Peeking inside one of the windows, Barney was totally unprepared for what he saw: six alien creatures.

In disbelief he jumped back in the car and tore off down the road. But he and Betty didn't get far. As they drove wildly through the night trying to get away, they began to hear an unusual beeping sound. Then their car started to vibrate. Finally a weird mist settled over them.

The next thing they remember is arriving at home. But when they checked their watches, they found that the trip had taken two hours longer than it should have. Somewhere on that lonely stretch of dark road the Hills had lost two hours of their lives.

What did it all mean? They weren't sure. They just wanted to forget the entire incident.

But it wasn't going to go away that easily.

A short time later both Betty and Barney began to have horrible dreams and various health problems. A Boston psychiatrist suggested they undergo regressive hypnosis in order to find out what happened on the night of September 19.

In completely independent sessions the Hills discovered that on that evening they'd been

UFO FACTOID
Two stars named Zeta 1 and Zeta 2 comprise a binary-star system known as Zeta Reticuli, home region of the aliens who abducted the Hills.

abducted by small machine-like alien beings, with huge hairless heads, large cat-like eyes, and gray skin.

Both had been given painful medical tests. During Betty's examination the aliens plunged a long needle through her navel to see if she was pregnant or not. This was years before we would discover a similar procedure, one used by gynecologists and that we now call "laparoscopy."

Before returning them to their car the aliens gave the Hills a "star map" showing where they had come from: Zeta Reticuli, a double-star system about forty light-years from Earth, located in the constellation Reticulum.

Barney passed away in 1969 and Betty died in 2004. But right up until the end of their lives neither one ever wavered from their original story, one that author John Fuller turned into a book called *The Interrupted Journey*.

THE ALIEN ABDUCTION OF BETTY ANDREASSON

Let's look at another example, one that has understandably gone down in the annals of UFO history.

> UFO FACTOID
> Betty Andreasson had one of the longest and most detailed alien abductions ever recorded, far too lengthy to include here.

Casebook Study 25: On the night of January 25, 1967, in South Ashburnham, Massachusetts, Betty Andreasson was in her kitchen cleaning up after dinner. Her husband was in the hospital and her seven children and her parents were in the living room watching TV.

For no apparent reason the lights in her house started blinking on and off, finally going out completely. Next, a flashing red-orange light suddenly poured through the kitchen window. When the family peered outside to see what was going on they saw four strange beings walking toward the house.

The creatures were four feet tall, had large pear-shaped heads, gray skin, and huge wrap-around eyes. Being religious, and knowing nothing about UFOs or aliens, Betty thought they were angels. They certainly behaved like angels: the creatures walked right through her door without opening it!

After that Betty and her family's memories go blank, and like the Hills, they thought nothing more of the event.

But ten years later, in 1977, Betty was persuaded to undergo hypnotic regression, which took her back to the night of January 25, 1967. It was then that Betty remembered the entire story.

After the little angel-like aliens had put Betty's family into suspended animation, they took her outside to a small spacecraft, which flew up and connected itself to a much larger ship. She was then placed on an operating table and given a series of painful medical tests.

The aliens then took Betty through a dark tunnel that ended in a large room. Here she was told that she had a special mission on Earth, one that would influence the whole world.

Afterwards she was taken back to her home and dropped off. She noticed that an alien had stayed behind to guard her family members, all who were still in suspended animation. The aliens then put the entire family to bed and left in their spacecraft.

Famed ufologist Raymond Fowler has written several books about "the Andreasson Affair," and to this day Betty continues to stand firmly behind her account.

THE ALIEN ABDUCTION OF TRAVIS WALTON

Here's a look at another famous abduction case.

Casebook Study 26: It was just another night after a hard day of cutting wood and clearing forests. Or so Travis Walton and his six male friends from Snowflake, Arizona, thought as they climbed into their pickup truck for the drive home. This particular evening, November 5, 1975, would turn out to be anything but "just another night."

> UFO FACTOID
> Travis Walton may have been targeted beforehand by his alien captors, who seem to have called him telepathically to the scene of his abduction.

The long ride was a routine one and they were tired. So no one paid much attention when off to the right one of them noticed a strange light in the woods nearby. As it became brighter and brighter they decided to drive over for a closer look.

Approaching the light they were startled to see an enormous spacecraft hovering near the tops of the trees. Oddly, Travis jumped out and ran toward the object as his friends yelled for him to come back to the truck.

All of the sudden a blue beam of light came down from the UFO. As it struck Travis, it lifted him in the air, then threw him onto the ground again. Seeing this the driver of the truck panicked, hit the accelerator pedal, and drove off as fast as he could.

The six men waited for ten minutes, sure that the UFO would have left the area by then. But when they returned to the scene not only was there no UFO, there was also no Travis Walton. He had completely disappeared.

Where had he gone?

This question wouldn't be answered until five days later, when Travis was found laying on a road miles away from where he was last seen in the woods. He was cold and unshaven, and had lost a lot of weight. He had no explanation for where he had been or what had happened to him. But under hypnotic regression Travis was able to recall many details of a terrifying ordeal.

> UFO FACTOID
> Travis Walton's abduction experience, witnessed by six other people, remains one of the most frightening, remarkable, compelling, and inexplicable in human history.

After his friends sped off he was beamed aboard the spacecraft and medically examined by a group of small gray beings with large domed heads and huge black eyes. He drifted in and out of consciousness, more scared than he'd ever been in his life.

The last thing he remembered was laying on the road, watching the UFO lift off and streak away into the sky.

Even though decades have passed since Travis' abduction, all seven men have continued to tell the same story, and have even passed several lie-detector tests. Travis wrote a book about his experience called *Fire in the Sky: The Walton Story*, and in 1993 a movie of the same name was made and released.

As we are about to see, however, humans are not the only creatures that aliens are abducting.

CHAPTER 4 AT A GLANCE

- Aliens have a secret agenda that may include: 1) the search for a new planet to colonize, and 2) the desire to create a new race of hybrid creatures that are half-human, half-alien, by mixing our DNA with theirs.
- When aliens take a human against her or his will it's called an "abduction."
- Humans who've been abducted are called "abductees" or "contactees."
- 4 million Americans say that they've been abducted by aliens.
- An abduction is a terrifying experience in which aliens kidnap a person from their home or car, take them into their spacecraft, and often perform painful medical experiments on them.
- The aliens block the memories of abductees so they can't remember exactly what happened.
- Many abductees experience what's called "missing time": a portion of time that's passed but can't be accounted for.
- Many abductees find strange marks on their bodies after their abduction, from bruises, abrasions, and cuts to incisions, scoop-marks, and triangular scars.
- Sometimes aliens "tag" their human victims by placing implants in their bodies. The implants are tracking devices that help the aliens keep an eye on the people they've abducted. These small instruments may also act as a medical evaluator, measuring an abductee's heart rate and other vital signs.
- After an abduction an abductee may experience many different things, from paranoia and health problems, to new psychic abilities and positive spiritual feelings.

5

CATTLE MUTES & MYSTERY CHOPPERS

"We have stacks of reports about flying saucers. We take them seriously when you consider we have lost many men and planes trying to intercept them." — General Benjamin Chidlaw, U.S. Air Defense Command, 1953

THE RIDDLE OF CATTLE MUTILATIONS

One of the eeriest, and scariest, aspects of the UFO phenomenon began in the U.S. in the fall of 1973.

It was in this year that farmers and ranchers across the Western and Midwestern states began to find that their cattle were being killed under extremely bizarre circumstances. Not by guns, or even by animal predators. But by medical procedures.

UFO FACTOID
Across the U.S., thousands of cattle have been mysteriously killed and horribly mutilated, always without leaving a single trace of blood, footprints, or tire tracks.

From California, Texas, South Dakota, Idaho, Montana, and Arizona, to Kansas, Wisconsin, Nebraska, Minnesota, and Iowa, dead cattle were discovered whose ears, eyes, tongues, lips, genitals, and tails had been sliced off or cut out. To make a strange situation even stranger, all of the blood had been drained from the animals. Yet not a drop of spilled blood was found in the area.

And, if this wasn't weird enough, there weren't any footprints or tire tracks anywhere near the carcasses. Dead cattle were even found in deep snow, freshly plowed fields, and mud holes. But still there were no signs of human presence around them. Nothing.

But there were a few clues to the mystery.

Some of the cattle had broken legs. Some were discovered high up in trees. Others were found with their horns stuck in the ground. It was as if the cattle had been magically lifted up into some kind of machine, operated on, then dropped from the air back onto the ground.

Adding to the grizzly enigma was the fact that the cattle were always taken and returned within sight of people's homes and farms, or near well-traveled roads. Yet no one had ever heard or seen anyone, or anything, come and go.

Odd lights, weird-looking helicopters, and even UFOs, were later seen overflying the areas where the cattle were found. But no one had ever been able to make any definite connection between these objects and the dead animals.

And these were not routine medical operations. They seemed to be nonhuman. Professional veterinarians examined the bodies of the dead cattle and found that whoever, or whatever, performed these operations was an extremely talented surgeon with skills that far surpass our own. The neat surgical cuts had not been made by a knife—as we would expect—but by a very delicate, very precise, very hot instrument. Like a laser beam.

These odd killings continue to occur, now all over the United States. And they're being reported in Canada, Western Europe, Central and South America, and the Canary Islands, now too.

And cattle are not the only creatures being mutilated. In New England small animals, such as rabbits and geese, have recently been found with strange surgical marks on their bodies.

> UFO FACTOID
> Skeptics assert that cattle mutilations are the result of natural predators, such as wolves and coyotes. However, the removal of organs from the dead cattle is always performed with neat, steady, careful, surgical precision, quite unlike the mindless tearing that's made by wild animals.

Welcome to the high strangeness realm of animal mutilations, or "mutes," as they're also called.

WHO ARE THE MUTILATORS?: TRYING TO EXPLAIN THE UNEXPLAINABLE

What kind of a person would want to do this? And why would they want to do it? What happens to the body parts that the mutilators so carefully remove? And what are these body parts being used for?

Again, we're trying to solve a dark mystery that has few clues, a sinister puzzle with too many missing pieces.

Mutologists, people who study animal mutilations, have many theories. Some believe that the cattle are simply dying from natural causes, such as disease and old age.

Others think that they're being killed and torn apart by predators, like bears, wolves, mountain lions, or coyotes.

Another group believes that the "mutilations" are simply the result of the animals' rotting flesh, which gives off gases that expand, sometimes tearing the skin or hide in nearly straight lines.

Still others maintain that cultists (a group of people who use animals in their religious rituals) are to blame.

But none of these suggestions even come close to explaining why nothing is ever seen or heard, why there aren't any tracks or footprints around the bodies, how the animals are lifted into the air, or how the highly skilled incisions in the cattle are made. Besides, professional investigators have never been able to find any evidence to support these hypotheses.

> UFO FACTOID
> A typical mutilated cow has had the skin around its lower jaw neatly cut and peeled away, something no animal is capable of doing. Mutologists believe this type of mutilation is done using a hot laser beam-like instrument. But our portable lasers require too much electricity and are not accurate enough to perform such long, arduous, and highly skilled operations out in the wild. Who, or what, is killing and carefully cutting up our cows then?

A much more reasonable explanation would be that a secret U.S. group is harvesting animal parts for some black project. They simply fly in over a farm at night in a helicopter. They choose a cow and hover over it, strap it into a harness, then hoist it up and operate on it. After taking the body parts they want, they drop it to the ground from the helicopter.

If only the answer was this simple.

Actually there are many problems with this theory. First, no one ever sees or hears the machines that take the cattle away. So the aircraft that are used must be completely silent. But as we've seen, humans don't have the technology to make silent flying machines yet.

Second, the incredibly precise surgical cuts seem to be made with a laser. But how could such a delicate operation be done in a hovering heavily vibrating craft like a helicopter?

Third, most lasers, and the machinery that goes with them, are large, bulky, and expensive to buy and operate. Even our newly developed portable lasers don't meet the requirements: not only can they not produce the precise incisions found in true mutilations, they also need tremendous amounts of electricity to operate.

The problem here is that in most cattle mutilations, a dozen or more organs are often neatly excised and removed—and this without spilling a drop of blood or leaving any tracks behind. All but the most close-minded skeptics agree that this is a process that would require hours to complete, if it were even possible to begin with.

At this point we must ask ourselves: even if we had an instrument capable of making surgical-like cuts in animal flesh, how could such an instrument be loaded on, and used from, a moving helicopter—especially in the middle of winter, the time period when so many cattle mutilations take place? And how could even the world's best surgeon operate under these harsh conditions, and without once being seen by anyone on the ground?

UFO FACTOID
Cattle are the most common mutilation victims. But other types of animals have been reported as well.

Fourth, why would a person want the ears, lips, eyes, tails, and reproductive organs from a common everyday animal like a cow in the first place?

And fifth, even if a person did find some use for these body parts, why would they go to all the trouble, danger, and expense of taking them in this way? Why not just buy them from a slaughterhouse for a few dollars?

In short, if our government wanted a pair of cow's eyes for some experiment, for example, there are much easier, faster, cheaper, safer, and simpler ways of procuring them.

THE UFO/CATTLE MUTILATION CONNECTION

Because none of these earthly theories come close to solving the mystery, we must turn to a more extreme possibility: UFOs and aliens.

As we learned from studying the alien agenda, some types of aliens are visiting Earth for the purpose of harvesting human DNA, or human genetic material. They seem to be combining it with their own in an attempt to create a new hybrid species that's half-alien, half-human. But what does this have to do with cattle mutilations?

UFO FACTOID
Black unmarked helicopters usually operate singly or in pairs, but squadrons of them have also been seen chasing or even accompanying UFOs.

What if the aliens can't get enough DNA from humans? Or what if it's too much trouble for them to abduct humans all the time? The answer is that they'd simply look for another source.

It just so happens that cattle DNA is very similar to human DNA. Knowing this, the aliens may be taking the DNA from cattle to make up for the difficulty of getting human DNA.

According to this theory the aliens would shape-shift their own spacecraft into something that closely resembles our helicopters. The helicopters would, of course, be:

1) silent, so no one can hear them
2) lightless, so no one can see them
3) unmarked (like UFOs themselves), so that no one can identify them, even if they're detected

They'd then fly quietly in over a cattle ranch at night and lift a cow up into the vehicle with a light beam (the same kind they use to abduct humans). After performing various highly advanced medical operations and tests on the animal, they'd remove some of its body parts and drop it back down to the ground.

While cattle abduction and mutilation is both illegal and frightening to the public, the military doesn't do anything to stop it. Why? Because of the Operation Majestic-12 (**MJ-12**) agreement, which allows aliens to abduct humans and animals in exchange for extraterrestrial technology. (More on this shortly.)

UFO FACTOID
The cattle mutilation phenomenon is a mystery that will not go away, no matter how hard authorities try to cover it up. Incidences of this freakish phenomenon are only increasing, and the government has run out of silly explanations.

This explanation may seem too incredible to be true. But it's the only one that answers nearly all of the perplexing questions about cattle mutilations. And besides, there are eyewitnesses to back it up.

Several people have actually seen UFOs in the process of abducting cattle. But since the aliens somehow blocked their memories, they could only recall these sightings while under hypnosis.

According to their accounts aliens used "pale yellow" beams of light to lift cattle up inside their spacecraft. Other hypnotized subjects say they saw aliens cutting up cattle and putting the body parts in huge vats or barrels. Are these stories for real? You decide.

REAL LIFE CASES OF CATTLE MUTILATION

Whoever it is that's taking and killing cattle one thing is certain: cattle mutilations themselves are very real. Here are a few cases from authentic police reports.

Casebook Study 27: From September through November 1975, numerous unmarked helicopters were spotted flying around the region of Baca County, Colorado. During this same period twelve cattle were found mutilated.

Casebook Study 28: On October 9, 1975, in Costilla County, Colorado, two police officers saw a helicopter land, then take off three minutes later. The next day two cattle were found mutilated in the area.

Casebook Study 29: On the night of November 3, 1975, a rancher in New Mexico saw several low-flying helicopters passing over his property. The next day he discovered that four of his cattle were missing.

Casebook Study 30: In early November 1975 hunters noticed a low-flying helicopter land in the area of Buffalo, Wyoming. Four calves were later found mutilated.

Casebook Study 31: Between October 30 and November 13, 1975, thirty odd looking helicopters and obvious UFOs were seen in the

UFO FACTOID
The American government openly makes black helicopters, such as the unmanned Northrop Grumman Ryan 379 VTUAV. But unlike true mystery helicopters, the 379 is not silent and is not flown over populated areas.

area of Union and Quay Counties, New Mexico. Four cattle mutilations took place during this same period. This case so baffled the **FAA** that they launched their own investigation in an attempt to find out what the unidentified aircraft were.

THE MYSTERY OF THE MYSTERY HELICOPTERS

As you read the many reports of cattle mutilations it's impossible to ignore the frequent mention of helicopters. And it's here that we find yet another bizarre phenomenon connected to UFOs and aliens.

Ufologists refer to these strange craft as "**mystery helicopters**," choppers that often show up flying around areas where UFOs have been sighted, or, as we've just seen, where cattle mutilations have occurred.

These spooky objects have been seen all over the world by thousands of people, including those who are fully trained in aircraft recognition, such as military personnel. But still no one has ever caught one, shot one down, or even been able to identify one. The reason? They're unlit and unmarked.

Mystery helicopters have no call letters, numbers, names, or symbols on them of any kind. So no one knows exactly who they belong to, who's flying them, where they come from, where they go, or why they're being flown.

Adding to their strangeness are the facts that they rarely use lights, even at night, and they're usually painted completely black (though some have been seen that are dark green or dark blue). Also, their windows are tinted so that it's impossible to see inside.

UFO FACTOID
Black mystery helicopters are not only unmarked with tinted windows, they also behave strangely and perform illegal maneuvers.

Mystery helicopters even behave bizarrely, usually flying illegally—and unsafely—low to the ground; performing unnecessary aggressive maneuvers (such as chasing or hovering over witnesses); and sometimes flying in huge squadrons. Normal helicopters don't do any of these things.

Witnesses report that some mystery helicopters even fly completely silently. Still others have seen dozens of weird lights flying around them.

Perhaps strangest of all is that mystery helicopters have been seen pacing leisurely alongside huge UFOs, like hummingbirds accompanying a jumbo jet.

EXPLORING THE NATURE OF THE MYSTERY HELICOPTERS

Many questions come to mind. What are the mystery helicopters? Where do they come from? Who's piloting them? What is their purpose?

Let's look at a few theories.

First, if we assume that they're real human-made helicopters, then we also have to assume that they're flying with the permission of someone at a very high level in the government. Why? Because they're intentionally being allowed to break the law. For the FAA strictly forbids any aircraft to fly without call letters, or some kind of identifying markings.

Maybe they're real helicopters then, working for a **black government** on **black projects**. Perhaps their main purpose is to monitor UFO activity.

> UFO FACTOID
> A Native American rock painting, made some 9,000 years ago, depicts a large glowing disk with portholes hovering over a group of people who are fleeing in terror. It appears to be an ancient picture of a UFO preparing to abduct someone. If not, what else could it be?

But is this really possible? It is. We've seen how governments have been hiding UFOs for decades behind a secret conspiracy of silence and denial.

The problem with this theory, however, is that mystery helicopters don't always appear or behave like actual human-made aircraft.

For example, sometimes they look like real helicopters but don't sound like them. Other times they sound like real helicopters but don't look like them. And at other times they don't look or sound like real helicopters but they act like them.

Another theory is that mystery helicopters are actually UFOs disguised to look like human-made terrestrial vehicles. This is also possible since, as we've learned, UFOs seem to be able to shape-shift their craft into looking like almost anything they want. This would certainly help explain why mystery helicopters look and behave so extraordinarily.

Reinforcing this explanation is the fact that real helicopters can be shot down. But not mystery helicopters, or real UFOs. During the Vietnam War hundreds of mystery helicopters were sighted by, and shot at by, American soldiers. But no wreckage of a mystery helicopter was ever found.

They were also chased by U.S. fighter jets, but none were ever caught. How is this possible? Jets can fly much faster than helicopters.

So what's behind mystery helicopters? They're obviously real: not only have they been painted on radar, but they've also been seen, heard, filmed, shot at, and chased by thousands of eyewitnesses.

The trouble is that mystery helicopters seem unreal. And because of this we really don't know what they are. Perhaps we can get a better idea by looking at a few actual cases involving these unearthly machines.

A REAL LIFE ENCOUNTER WITH MYSTERY HELICOPTERS

As we've seen, most mystery helicopter sightings seem to be connected with cattle mutilations. But not all. Here's one of the more dramatic examples of the connection between mystery helicopters and UFOs.

Casebook Study 32: On the night of December 29, 1980, not far from Dayton, Texas, Betty Cash, and Vickie Landrum and her seven-year-old grandson, Colby Landrum, were casually driving home.

Traveling along they noticed an extremely bright light ahead. As they got closer they couldn't believe their eyes. They were looking at a large diamond-shaped UFO hovering in the air just above the road. As it gently bobbed up and down, beeping sounds were coming from it and bursts of flame shot out of the bottom.

The light from the craft was so intense that it hurt their eyes. And even though it was wintertime the inside of the car got so hot they had to turn the air conditioner on.

UFO FACTOID
What did Betty Cash and the Landrums witness that winter night? And why were they so severely injured? Skeptics say that it was all the result of an American military project. But if so, why, when asked, did the U.S. government deny any association with the incident, *and* refuse to help pay for the families' medical costs?

UFO FACTOID
The Cash/Landrum Incident continues to defy all explanations. Except for the UFO one.

As they watched in disbelief the unearthly object lifted up into the sky, moved away, and eventually disappeared from view. But the show wasn't over yet. As the three drove off to continue their journey home they saw the UFO reappear. But this time it was surrounded by twenty black unmarked helicopters. As the strangely behaving aircraft sped overhead they swept the area with huge searchlights, as if they were looking for something.

In the following days strange unmarked trucks and work crews appeared at the site and, mysteriously, resurfaced the road, perhaps a governmental effort to prevent investigators from gathering any lingering evidence, radioactive or otherwise. But the incident would not go away, researchers were not discouraged, and no civilian, particularly Betty, Vickie, and Colby, was about to forget about it.

Indeed, as the weeks and months went by all three were constantly and painfully reminded of the sighting: they had begun experiencing serious health problems, from headaches, diarrhea, and nausea, to weight loss, swollen eyes, and hair loss. Red marks broke out on their skin, later turning into blisters. The diagnosis? Radiation poisoning.

Sympathetic investigators called all of the nearby airports and military bases to report what Betty, Vickie, and Colby had seen, and to ask about the helicopters. But no matter who they called, everyone gave them the same reply: "Not only did we not have any helicopters in the sky that night, but we don't even have that many helicopters."

What were these unusual aircraft that so terrified Betty, Vickie, and Colby? Obviously they were mystery helicopters. But who or what is operating them? As we're about to see, it may be a group of governmental organizations so secret that our own world leaders aren't even aware of them. Or they may be alien spacecraft.

CHAPTER 5 AT A GLANCE

- In 1973 American ranchers began to notice that their cattle were dying from what looked like horrible medical procedures performed by laser beams.

- No tracks or footprints were found around the dead bodies, but strange lights, UFOs, and unmarked helicopters were seen in the area.
- These weird deaths gave rise to a new science: mutology, the scientific study of animal mutilations.
- Animal mutilations are now occurring all over the world.
- The identity and purpose of the mutilators remains unknown, though there are many theories: 1) natural causes, like disease; 2) predators, like wolves; 3) cultists; 4) a black government project; 5) aliens looking for DNA to replace human DNA.
- Mystery helicopters are aircraft that look, sound, and act similar to real helicopters.
- Mystery helicopters operate illegally by flying without identification numbers, names, or symbols.
- They're almost always unmarked, painted black, and operate without lights.
- They often fly silently and their windows are tinted a dark color.
- They fly low to the ground, behave aggressively, and can fly alone or in large formations.
- Mystery helicopters have been seen flying alongside and around UFOs.
- They may be the secret aircraft of a black government whose purpose is to explore the UFO phenomenon.
- Mystery helicopters could also be UFOs that are disguising themselves to look like human-made helicopters.
- The Cash-Landrum Incident resulted in serious health problems for the three eyewitnesses, but no interest in the sighting was ever shown, publically at least, by the government.

6

LIES, DENIALS, & SECRETS: THE WORLDWIDE UFO COVERUP

"From available information, the UFO phenomenon appears to have been global in nature for almost 50,000 years. . . . This leaves us with the unpleasant possibility of alien visitors to our planet, or at least of alien controlled UFOs." — USAF Academy Textbook, Department of Physics, 1969

BLACK GOVERNMENTS, BLACK PROJECTS, & BLACK BUDGETS

One of the more shocking aspects of the UFO phenomenon is that governments around the world continue to pretend that UFOs don't exist, even though they know full well that they do. They seem to want the public to think that UFOs are an illusion, while behind the scenes they take them very seriously, even creating entire committees and departments that do nothing else but study, report on, and film UFOs.

UFO FACTOID
It's difficult for the government to hide the reality of UFOs when they fly right over our cities in broad daylight, sometimes before millions of eyewitnesses.

Some agencies, like the CIA, for example, secretly monitor every UFO report from every country on Earth. And yet at the same time, it tells the public that it has "absolutely no interest in UFOs."

Proof of the agency's duplicity comes from individuals like former CIA official Victor Marchetti, who now openly admits that the U.S. government indeed has had interaction with aliens. However, in agreement with other governments around the world, this information is purposefully being kept from the public.

What's going on here?

In many countries, such as the U.S. and the U.K., there are smaller secret governments, called **"black governments"** (also known as **"para-governments"** or **shadow governments**), operating deep within the main governments. Black governments are so secret that even the American president and the British prime minister probably don't know anything about them. In fact, these organizations are so covert that they're classified as "above Top Secret."

UFO FACTOID
The governments of the world tell us that "UFOs don't exist," then classify them "above Top Secret." Many of us would like to know why.

What is their purpose? Black governments were created specifically to work on what are called **black projects**: highly secret programs known only to a few carefully selected individuals or groups. Examples of black projects would include the development of new top secret spacecraft and military weapons.

Where does the money come from to work on black projects? Black projects are funded by **black budgets**, huge sums of money secretly supplied by the main government. In 1993, for instance, the U.S. Department of Defense alone requested 16 billion dollars for black projects. We only speculate as to what other governmental agencies are involved. Probably many of them are unknown to the public.

THE GOVERNMENT'S BIGGEST SECRET: THE UFO COVERUP

Much of the money from black budgets goes to the U.S. government's most important—and most secret—black project of all: the study, and the coverup, of UFOs and aliens.

Government documents and eyewitness testimonies reveal that while they publicly deny interest in UFOs, black governments have actually captured UFOs and living aliens, and are holding them right now in unknown underground military installations.

The governments of the world are not only covering up the facts about UFOs, they're also trying to prevent us from ever knowing the truth. This is why we call this official deception "the UFO Coverup." And it's also why one of France's best science researchers, Fernand

Lagarde, calls the government officials who head the worldwide coverup "the Masters of Silence."

Just how do the Masters of Silence conceal the truth about aliens and UFOs?

First, they follow a simple formula to discourage questions on the topic: "deny everything!"

Next they put out **disinformation**. This means that they intentionally create and leak bad information, pretending that it's good information. This could include making fake documents, tapes, digital recordings, and photographs, spreading false rumors, ridiculing UFO witnesses, infiltrating UFO organizations, hoaxing UFO sightings, planting false UFO stories in newspapers and magazines, and even discrediting military pilots who've seen UFOs.

> UFO FACTOID
> The way our governments handle the topic of UFOs is guaranteed to create suspicion, fear, anger, and charges of conspiracy and coverup.

The Masters of Silence then ridicule those people who come to accept the very deceptions that they create. The purpose? To mislead, confuse, and deceive, in an attempt to cover up the truth about UFOs.

This is not science fiction. Some government agencies, such as the **KGB**'s notorious "Department D," do nothing else but spread disinformation—often about UFOs.

Some types of disinformation are very hard to spot. For instance, various space research organizations tell us that though there must be other living beings in our Universe, they probably live too far away to contact us let alone physically visit us. According to this line of thinking, even if we actually make contact with aliens someday it will be by way of **radioastronomy**, not face-to-face.

To reinforce this deception the Masters of Silence create projects like SETI (the "Search for ExtraTerrestrial Intelligence"), which explores the skies using special listening devices in an attempt to pick up signals from alien civilizations living in outer space.

Projects like **SETI**, however, are a waste of time and money. Why? Because, as any reasoning person can see, alien civilizations are already visiting Earth. It's just that the government doesn't want us to know this, which is why SETI was created to begin with. This makes disinformation programs like SETI ideal for covering up the truth about UFOs.

WHY GOVERNMENTS ARE COVERING UP UFOS

But why would the world's governments want to keep the reality of UFOs a secret? There are several possible reasons.

UFO FACTOID
Are the world's government powerless against alien technology? The facts point to only one answer: yes.

1. *The Embarrassment Factor*: The government and the military would be deeply embarrassed if the public found out that they can't protect us from alien invaders. UFOs simply travel too fast for jets to catch them, and our guns and missiles aren't quick enough to shoot them down. UFOs have even been known to dodge bullets (which travel at many thousands of miles an hour).

Also, how can our military be expected to hit an object that has the ability to change its shape and size, or even disappear into thin air? And let's not forget that UFOs have the power to turn off our electrical, radar, and weapons systems.

The truth is that no military organization on Earth has any real defense against UFOs—as the UFOs themselves have shown us time and time again. This is a terrifying reality that no government wants its people to know.

2. *The Scientific Pride Factor*: If the facts about UFOs and aliens were made public, nearly every scientific idea ever created would instantly become obsolete. And every science book that's ever been penned would have to be thrown out and rewritten from scratch.

In other words, the reality of UFOs would make all of our current scientific knowledge irrelevant, and thus unusable.

Scientists understandably take great pride in their work, ideas, and theories, and they deserve our respect. However, many have overly inflated egos and use science to achieve personal glory rather than to help humanity progress. This type of scientist wouldn't be very happy about having everything he or she has created—not to mention everything they personally believe in—thrown in the trash overnight.

UFO FACTOID
The U.S. government thinks that the public is not ready to hear the truth about unidentified flying objects and ETs. However, the majority of Americans already believe in, and accept, the reality of the worldwide UFO phenomenon. Why keep hiding it then?

3. *The Mass Hysteria Factor*: Many governmental people don't want the public to know the truth about UFOs because they think that human society would collapse.

According to them, mass hysteria would break out if we were to learn the truth. In the panic, fear, and chaos that would follow, they assume that not only our scientific ideas, but also our religions, our economies, our legal systems, our social systems, even our governments themselves, would become meaningless, causing total disruption to human life.

But those who believe this are wrong.

First, the military and government people who do know the truth about UFOs and aliens haven't become mentally deranged, self-destructive, psychopathic, or suicidal.

Second, most private citizens already accept the fact that alien beings are visiting Earth, and they haven't gone mad either. All in all, human society is still running quite smoothly.

Third, many groups, organizations, and religions have always believed in the reality of UFOs and aliens. Some, like the Mormons, or Church of Jesus Christ of Latter-day Saints (LDS), actually teach that there are many other lifeforms in our Universe and that Mormons themselves can become "gods" and "goddesses" of other planets after death.

The Catholic Church too openly acknowledges the existence of UFOs and aliens. One Vatican theologian, Monsignor Corrado Balducci, states that "extraterrestrial contact is real," noting that the Church is monitoring the UFO phenomenon carefully, preparing for the day when the governments of the world finally announce the truth.

4. *The Technology Factor*: Perhaps the main reason UFOs are being covered up is because the world's governments are trying to duplicate UFO technology (using captured and crashed UFOs).

By keeping its citizens, and other governments, from knowing what they're developing, a government would have a great technological advantage over other countries and its enemies.

Imagine being the leader of a nation. Now imagine that you possess the superior technology of aliens, technology so amazing that you could lead the rest of the world both militarily and economically. Wouldn't you want to keep this a secret too?

CRASHED SAUCERS & WEATHER BALLOONS: THE ROSWELL INCIDENT

We don't know exactly when the U.S. government first became aware of the reality of UFOs and aliens, but it couldn't have happened any later than July 2, 1947. It was on this day, in Roswell, New Mexico, that visitors from outer space definitely came into contact with the U.S. government.

> UFO FACTOID
> Some religions, like the Mother-Goddess-worshiping Mormons, actually teach that there are other lifeforms in the Universe. Catholicism also accepts the fact that we are not alone, even stating that we should ready ourselves for the day when our governments acknowledge the worldwide UFO coverup.

Casebook Study 33: Roswell was once the site of the military's atomic and research and development facilities. Here the U.S. government tested its missiles and housed the world's only atomic bomber squadron. With the U.S. now having the capability to make nuclear weapons we can be sure that alien beings would eventually take an interest in the area. And take an interest they did.

On July 2, as an extraterrestrial spacecraft flew over Roswell to investigate the installation, something went terribly wrong: the UFO, along with its three alien occupants, crashed into the ground, creating a large red glow in the night sky and leaving a trail of debris across the New Mexico desert almost a mile long.

Local ranchers, the first on the scene, surveyed the crash site and found a strange aluminum-like material that couldn't be cut, scratched,

> UFO FACTOID
> Currently the Roswell Incident is America's greatest UFO event. Some go even further, saying that it's the world's most spectacular extraterrestrial episode ever.

or torn, and that unfurled by itself when rolled into a ball. They also discovered bizarre markings, a form of writing similar to Egyptian hieroglyphs, on some of the metal I-beams taken from the crashed UFO.

When the military heard about the accident they rushed to the scene and quickly took both the wreckage and the aliens away, locking them in a top secret airplane hangar. The military then reported to the newspapers exactly what it had found: a crashed UFO. The original July 8th headline actually read:

> RAAF CAPTURES FLYING SAUCER ON RANCH IN ROSWELL REGION: No Details of Flying Disk are Revealed

The article went on to say:

> The many rumors regarding the flying disk became a reality yesterday when the intelligence office of the 509th Bomb Group of the Eighth Air Force, Roswell Army Air Field, was fortunate enough to gain possession of a disk through the cooperation of one of the local ranchers and the sheriff's office of Chaves County.

The next day, however, realizing the seriousness of the situation, the military changed its mind. Now it decided to call the weird spacecraft a "weather research balloon," after which it replaced the real UFO wreckage with fake wreckage from a real weather balloon. The original (UFO) wreckage was then quickly and secretly trucked out, finally ending up in a base hangar at Wright Field in Dayton, Ohio.

FBI Director, J. Edgar Hoover, was not happy, sending out an angry memo to his cabinet that read: "I must insist upon full access to discs recovered. . . . The Army grabbed one and would not let us have it for cursory examination."

The American public was not pleased either, completely rejecting the military's explanation for what crashed in the Roswell desert that night. And for good reason: if the downed object was nothing more than a weather balloon, why cover it up behind a huge shroud of secrecy?

The public knew it was a UFO and dozens of eyewitnesses later came forward to verify this sentiment.

Local police, for example, recalled seeing the crash sight, with two dead aliens laying amid the wreckage—and one still very much alive walking around in a daze. Nurses described the operating room at the military base and the dead aliens who were medically examined and filmed there. Government employees told how they later studied the ruined saucer in an attempt to learn how it operated.

> UFO FACTOID
> Eyewitnesses at Roswell say that the dead aliens found at the crash site were later examined by military doctors. Some even claim to have photos and videotape of the alien autopsies. All were threatened to keep quiet.

At the same time, workers who helped cleaned up the wreckage at the crash site were told that it was their "patriotic duty to remain silent." And radio station managers in the area were warned that they'd be shut down if they tried to broadcast anything about the story.

Additionally, dozens of local Roswell residents who also witnessed the incident said the military promised violence against them and their families if they ever spoke to anyone about what they saw. A government official told one woman that she would be taken out to the desert and "never heard from again" if she ever revealed what she had seen at Roswell. When she asked why, he replied angrily: "You don't have a need to know!"

"A POLICY OF DENIAL" WITHIN A "CONSPIRACY OF SILENCE"

From that day forward the U.S. government has followed a "policy of denial" when it comes to UFOs. Because of this the public has assumed that behind the government's unconvincing explanation of Roswell there is a major "conspiracy of silence" to hide the truth about UFOs and aliens from the world. Indeed, the very fact that the government refuses to even be questioned on the topic tells us that they're hiding something of great importance.

And so, despite all of the eyewitness testimony of what actually happened there, to this day the U.S. government continues to conceal the truth about what it recovered at Roswell.

UFO FACTOID
The American government's *The GAO Report on Roswell* was a pitiful and transparent attempt to end all discussion of the subject. It utterly failed. The controversy, intrigue, and debate continues, stronger than ever.

The "final" official U.S. governmental denial came in 1995 when the **GAO** published a book about the Roswell Incident in an effort to appease the public and permanently "close the book" on the case. Entitled *The GAO Report on Roswell*, it found, not surprisingly, that all information (papers, documents, files, tapes, photos, etc.) concerning Roswell had disappeared, and concluded that "there is no evidence that a UFO or aliens were ever found."

This wouldn't be the last of such underground plots to conceal the facts. The U.S. government has engaged in many coverups over the years, from the Agent Orange disaster to the Tuskegee experiments, from the Watergate Break-in to Gulf War Syndrome. Indeed, the Roswell coverup itself may have brought to light one of the government's more fascinating supersecret black projects to hide UFOs.

Called "Operation Majestic-12," or **MJ-12**, it began in the late 1940s under President Harry Truman. Recently discovered government documents show that around this time a panel of twelve men were assigned to make a deal with the aliens: in exchange for UFO technology the government would allow the aliens to take humans and animals (such as cattle) and perform experiments on them.

The twelve men (all now deceased) who ran MJ-12 are:

1. Admiral Roscoe H. Hillenkoetter
2. Dr. Vannevar Bush
3. Secretary James V. Forrestal
4. General Nathan F. Twining
5. General Hoyt S. Vandenberg
6. Dr. Detlev Bronk
7. Dr. Jerome Hunsaker
8. Mr. Sidney W. Souers
9. Mr. Gordon Gray
10. Dr. Donald Menzel

11. General Robert M. Montague
12. Dr. Lloyd V. Berkner

If this theory, known as "the Majestic-12 Conspiracy," is true, then it really is the aliens who are in charge of humanity's future, not we ourselves.

> UFO FACTOID
> American president, Harry Truman, initiated the covert UFO organization called MJ-12 in the 1940s. This single secret act would change the course of human history.

CRASHED UFOS WITH STRANGE WRITINGS

Government coverups of UFOs and aliens didn't stop after Roswell of course. Estimates are that as many as one or two crashed saucers (and their occupants) are secretly recovered by the military each year somewhere in the world.

Many, like the one picked up at Roswell, are found to have strange markings, symbols, or writing on them. One famous example of this type of craft dates from December 9, 1965, when an acorn shaped UFO was spotted by dozens of people as it flew over six states and part of Canada. Finally, it caught fire, made a loop in the air (proving that it was under intelligent control and not a meteor, as skeptics assert), and crashed into a forest near Kecksburg, Pennsylvania. Eyewitnesses reported that "it was not of this world," and that it had hieroglyphic-like writing along its rim.

But before these, or any other part of the UFO, could be examined by civilians, it was quickly placed under a tarp and hauled away by military personnel on the back of a flatbed truck. As ususal, the government subsequently covered up the entire incident.

RED-EYED ALIENS & THE COVERUP AT VARGINHA

A spectacular and recent example of another crashed UFO, this one complete with prowling, menacing aliens, occurred in 1996 in Brazil.

Casebook Study 34: At 1:00 A.M. on January 20, people in the city of Varginha noticed a bright light overhead in the sky. Looking up they watched for forty-five minutes as a bright submarine-shaped UFO hovered in the air, then moved off into the distance.

The mysterious craft must have crashed in a forest nearby not long after, because the next morning a number of bewildered and frightened alien beings were seen wandering around town.

UFO FACTOID
Shortly after a UFO was spotted over the Brazilian town of Varginha in January 1996, a group of bizarre creatures was seen prowling the streets.

The first was spotted by three girls who were walking past a vacant yard. There they noticed a truly bizarre creature crouching by a wall. It had large glowing red eyes, bulging veins, a small mouth, green leathery skin, and huge three-toed feet.

Thinking it was "the Devil," the girls ran home and called the fire department. A few minutes later they saw several firefighters carrying the struggling creature away in a net.

Townspeople soon noticed military personnel entering the woods that surrounded Varginha. Witnesses report that they heard three shots, then saw soldiers walking quickly back out of the forest carrying two sealed canvas bags. Whatever was inside one of the bags was moving.

But not all of the aliens were caught.

Months later, on April 21, keepers at the local zoo were trying to figure out why all of the animals were suddenly so agitated. The answer wasn't long in coming.

A woman sitting on the zoo's veranda was startled to see that one of the red-eyed creatures had entered the park and was leaning on a fence staring at her. It stood there for seven long minutes as she sat frozen in her seat. Shortly after this episode many of the zoo animals died of unknown causes.

Ufologists and researchers quickly discovered that at least seven aliens had been seen since the night of January 20, 1996, and that two of them had been captured alive and taken to a local hospital for observation. The dead aliens had been flown out to the University of Campinas in the state of Sao Paulo. Here they were placed in a secret underground laboratory and examined.

UFO FACTOID
The Varginha Incident was covered up by the Brazilian government, just like Roswell was covered up by the U.S. government.

Ufologists also found out that one of the police officers who had helped pursue and capture one of the aliens later died as a result of injuries he had received from the creature.

Weeks after the incident eyewitnesses were visited by strange men who threatened them if they told anyone what they'd seen. As a reward for following their orders the witnesses were offered money and relocation.

When the firefighters and military personnel who were involved were later questioned by concerned civilians, they refused to talk, even denying that UFOs or aliens had been seen in the area. The Varginha Coverup was already in full swing.

> UFO FACTOID
> My cousin, Sir Winston Churchill of Great Britain, was very interested in UFOs and believed that they posed a potential threat to national security.

ABOVE TOP SECRET

Ever since Roswell, governments around the world have continued to show an interest in UFOs, despite the fact that they keep telling the public that there's no such thing. Even some of the people who work for the government are unhappy with the way their employer is concealing the truth about UFOs.

In July 1952, for instance, the United Kingdom's Prime Minister, Winston Churchill, sent off an angry letter to his Secretary of State. Feeling he'd been left out of the UFO information loop, Churchill wrote:

> What does all this stuff about flying saucers amount to? What can it mean? Let me have a report at your convenience.

The 1965 U.S. Speaker of the House of Representatives, John W. McCormack, also complained about the coverup, saying that the Air Force was withholding information about unidentified flying objects while trying to pretend that they don't exist. How can we ignore the many reported sightings made by "unimpeachable sources"?, he asked flatly.

In the 1970s American Senator Barry Goldwater tried repeatedly to obtain secret documents which showed that the government was doing research on UFOs. But in every instance he was denied. The explanation? "UFOs are classified above top secret."

Goldwater was understandably frustrated, for both military and commercial pilots were telling him that UFOs were coming right up to their aircraft, close enough for the pilots to see every detail of the otherworldly objects. Obviously the government is investigating UFOs, the senator complained, so why are they hiding the fact from us?

UFO FACTOID
At least 85 percent of Americans believe in the reality of UFOs and that there was a governmental coverup at Roswell.

Afraid that the truth will one day come out, it's no wonder that U.S. military authorities prefer to explain UFOs away as "misidentification," or as "unusual weather patterns," "space debris," or "meteor showers."

But who ever heard of a cloud, a burned out rocket, or a frozen chunk of mud that could change shape, size, and color, hover in midair, reverse itself, shoot straight up, zoom across the sky at 50,000 mph, and turn a ninety-degree angle without slowing down?

What intelligent person could ever take the military or the government's explanations seriously? Not very many: 65 percent of Americans believe that a UFO crashed near Roswell, while 85 percent believe that their government is lying about UFOs.

AREA 51: THE AMERICAN UFO TEST BASE THAT "DOESN'T EXIST"

The problem for the world's governments is that their own records, documents, and employee reports show that UFOs are real, that they've definitely known about UFOs since at least 1947, and that they're constantly working on recovered alien spacecraft.

For example, we have the testimony of several workers from "Dreamland," or Area 51—a top secret U.S. military base in Groom Lake, Nevada—who've come forward with amazing stories about black programs pertaining to UFOs and the government.

UFO FACTOID
Area 51 may be responsible for the large number of planes that have disappeared in what is known as the "Nevada Triangle."

UFO FACTOID
Anyone coming near Area 51 can't help but see the large warning signs posted around its perimeter, threatening the use of "deadly force" against trespassers. How many innocent citizens have been killed trying to uncover Area 51's mysteries? No one will ever know because any deaths that have occurred were quickly hidden beneath a mountain of bureaucratic secrecy.

According to them, the U.S. military is back- or reverse-engineering a number of captured and crashed UFOs that it keeps in underground chambers at Area 51.

The purpose? The military is taking these UFOs apart piece by piece in order to learn how alien spacecraft are made.

The reason? They hope to use this same technology to build their own secret aircraft someday.

The military, however, may have already succeeded. According to some sources the Air Force's stealth aircraft—the B-2 Spirit, the Lockheed U-2, the F-22 Raptor, the F-35 Lightning II, the SR-71 Blackbird, the F-117 Nighthawk, and the Aurora—were all developed using alien UFO technology. Unlike earlier aircraft, these sleek planes and jets (all tested in total secrecy at Area 51) are invisible to radar. They even look a bit "alien."

Others claim that improvements in lasers, night vision equipment, fiber optics, Kevlar, LED lights, and even the integrated circuit (used in computers), were brought about by using back-engineered UFO technology.

Eyewitnesses have climbed the mountains around Area 51 and videoed mysterious glowing star-like lights that dance, hover, and streak across the sky, even performing instant ninety-degree turns. Former employees tell us that these lights are actual alien UFOs that the military is trying to learn to fly.

Satellite photos of Groom Lake show a huge complex of assorted buildings such as warehouses and barracks, along with radar equipment, test planes, and the longest paved runway and the biggest airplane hangar in the world.

UFO FACTOID
Throughout the years many government employees have been aware of the worldwide UFO coverup but were powerless to do anything about it. Threatened with fines, job termination, or worse, they were forced to remain silent. It was the passage of America's FOIA in 1966 that finally allowed ufologists to begin to expose the truth.

Still, the government says that "Area 51 doesn't exist," and to this day the super secret base cannot be found on any map of Nevada. In other words, the U.S. government is even trying to cover up the site where they're covering up captured UFOs!

SECRET GOVERNMENT DOCUMENTS REVEAL THE TRUTH

Obviously the government isn't telling us the whole story because, until 1976, it was still stating that it had no interest in UFOs. And yet when U.S. Navy physicist Dr. Bruce Maccabee filed a **FOIA** request with the FBI that year, he received over 1,000 pages of UFO-related documents.

> UFO FACTOID
> Despite their best efforts to conceal the fact it's more than obvious that the world's governments are highly interested in UFOs and aliens.

As early as 1947, in a top secret document, the U.S. Air Command admitted that "the [UFO] phenomenon is something real and not visionary or fictitious."

A top secret Canadian government memo dated November 21, 1950, reads, in part, that:

> 1) The matter [of UFOs and alien beings] is the most highly classified subject in the United States Government, rating higher even than the H-bomb.
>
> 2) Flying saucers exist.
>
> 3) . . . concentrated effort is being made by a small group [to investigate UFOs and extraterrestrial beings—that is, MJ-12] . . .
>
> 4) The entire matter [of UFOs and aliens] is considered by the United States authorities to be of tremendous significance.

Another top secret U.S. document, this one from the **ATIC**, is entitled "Estimate of the Situation." Dated August 5, 1948, it reveals that government UFO research groups had, by this time, determined that UFOs are not from Earth, but instead originate somewhere in space. Shortly after its release government officials quickly ordered that the original document, and all copies, be burned and discarded.

After the passage of the **Freedom of Information Act (FOIA)** in 1966, the UFO Coverup really began to blow wide open. For the first time this gave ufologists, like Dr. Maccabee and other private citizens, the right to ask the government for copies of any and all classified documents it possessed that were related to UFOs. Even ufologists weren't prepared for the shocking information they found in these papers.

One 1947 document discovered through the FOIA, an FBI/Army Intelligence Report, plainly revealed what the government has known all along:

> The flying saucer situation is not imaginary . . . something is flying around.

Government documents from other countries also reveal a UFO coverup. On October 20, 1976, for instance, Spain's Air Ministry released 300 pages of its UFO files to the public. After reading them one alarmed researcher said that it was obvious that UFOs are real and that the national governments of the world are "deeply concerned" about them.

EIGHT UFO QUESTIONS TO ASK YOUR GOVERNMENT

If you're like this researcher, you too are deeply interested in learning the whole story.

If so, why not write to one of the political officials who represent your city, county, state, province, or nation? Ask them the following eight questions:

> UFO FACTOID
> Our governments repeatedly tell us that "there are no such things as UFOs," while actively investigating them behind the scenes. Why? And if UFOs aren't real or aren't a threat, why refuse to either discuss them or release documentation about them?

1. Why is our government wasting time debunking UFOs, something that it says it has no interest in?

2. How can the government claim that UFOs don't exist while labeling them "unidentified flying objects"? Even though they can't be

identified the use of this term shows that the government still admits that they're real objects. Therefore they must exist.

3. If UFOs aren't real why do the Air Force Intelligence Manuals of most countries contain drawings of UFOs that are meant to help pilots in identifying aerial craft?

4. If believing in UFOs is "silly" why did the Director General of the USAF issue a document called "UFOs Are Serious Business" to all Air Base Commanders on December 24, 1959?

5. Why does this document instruct pilots on how to collect samples of UFOs if UFOs aren't real to begin with?

6. If UFOs are not a threat and are imaginary anyway, why do so many of the UFO-related documents that are eventually released by the government have words, names, sentences, paragraphs, and even entire pages blacked out? What is there to hide if UFOs aren't real, or if they're just meteors, space garbage, mirages, Saint Elmo's Fire, ball lightning, Chinese fire lanterns, or weather balloons?

UFO FACTOID
Your government believes in UFOs. But it doesn't want you to.

7. If there are no such things as UFOs, why refuse to release the thousands of UFO-related documents, radar records, audio tapes, photographs, and films that you have on file?

8. And last but most importantly, if there are no such things as UFOs, why bother creating a complex, time-consuming, expensive, worldwide coverup in an attempt to hide them?

Why indeed. Especially when we have the sworn testimony of such distinguished people as the Canadian government's senior radio scientist, Wilbert B. Smith, who, in 1959, wrote that there is no government on Earth that has not been made officially aware of UFOs and aliens.

A clue to the worldwide coverup of UFOs can be found in the humble wheatfield, the topic of our next chapter.

CHAPTER 6 AT A GLANCE

- The governments of the world are engaging in a massive coverup of the truth about UFOs and aliens.
- Many governments have black governments that are working on black projects related to UFOs. These black programs are funded by black budgets.
- The worldwide UFO coverup is maintained through the spread of disinformation, which includes hoaxing UFO sightings, ridiculing UFO witnesses, infiltrating UFO groups, and planting false magazine and newspaper stories.
- The governments of the world have many reasons for keeping the truth about UFOs from the public: 1) they're embarrassed by the fact that they can't protect their citizens from them; 2) every scientific principle would have to be thrown out and rewritten; 3) we the public might panic, causing worldwide social, religious, political, and economic disruption; 4) they're duplicating alien technology from crashed UFOs, which will give them a military and industrial advantage over other nations.
- On July 2, 1947, a real flying saucer crashed near Roswell, New Mexico, but this fact was concealed by the government.
- The vast majority of Americans believe in UFOs and that there was a coverup at Roswell.
- When it comes to UFOs governments are engaging in a "policy of denial" and a "conspiracy of silence."
- Area 51 is a top secret military base—where captured UFOs are being back-engineered—but which the government claims "doesn't exist."
- Secret government documents, released through the FOIA, show without any doubt that the U.S. government has known about UFOs since at least 1947, and that it admits that they're real.
- If our governments believe in UFOs, why shouldn't we?

- We have many questions for our governments. For example, why create a massive and expensive coverup if UFOs don't exist?
- Though the U.S. government says "UFOs are not a threat to humanity," it lists them "above Top Secret," a security status so high that not even the president of the United States has access to this information.

7

VISITORS IN OUR FIELDS: CROP CIRCLES

"The phenomenon reported is something real and not visionary or fictitious." — General Nathan Twining, Chairman, Joint Chiefs of Staff, 1947

CROP CIRCLES: AN UNSOLVED ENIGMA

On August 15, 1980, an article appeared in the British newspaper, *Wiltshire Times*, describing a bizarre sixty-foot flattened circle of grain on a farm near Bratton, Wiltshire, England. This article marked the beginning of **cereology**: the scientific study of UGMs ("unusual ground markings"), or crop circles, as they're more popularly known.

Since that time thousands of other crop circles have been documented and photographed all over the world, from the United States and Canada, to Japan and Australia. Not only have the numbers of crop circles increased since the 1980s, but so has the artistic complexity of the circles.

> UFO FACTOID
> Crop circles range from tiny to enormous, from simple to complex, from humorous to serious.

What began as simple single circles have today evolved into a wide variety of shapes and sizes that include long straight lines, huge dumbbells, rung ladders, tiny triangles, enormous squares, giant spirals, medium-sized ovals, and even mathematical formulas. The largest crop circle ever recorded was almost a half mile long, the smallest a mere eight inches in diameter.

These strange Earth mysteries seem to appear primarily in the spring and summer and near the sacred sites of prehistoric and ancient societies. The stalks of the grain—whether they're wheat, oats, barley, or rye—are bent over without any damage, in either a clockwise or a counter-clockwise direction. Sometimes the plant stalks are braided, other times different layers of the stalks are laid down in different directions. But mysteriously in no case are they ever broken.

UFO FACTOID
Skeptics assert that all crop circles are human-made. Yet eyewitnesses have been reporting self-generating crop circles for hundreds of years.

FAIRY RINGS & SAUCER NESTS

No one knows when crop circles first began appearing, though there are reports of self-forming circular patterns emerging in fields dating from as early as the year 1590.

No one knows for sure how crop circles are made either. Many theories have been put forward. Some say that they're weather-based and that they're formed by a "plasma vortex": a ring of electrically charged air that creates a downward pushing whirlwind.

Others say that crop circles are constructed by fairies and elves dancing in a circle, which is why some people call them "fairy rings."

Yet another theory is that they're made by the Earth herself; that is, by the Great Earth-Mother-Goddess known as Gaia. As such, crop circles would have played an important role in ancient Goddess-worship.

Skeptics and debunkers, of course, claim that all crop circles are hoaxes made by humans.

But according to many **cereologists**, by far the most reasonable explanation is what's called the **TIF**, or "the Theory of Intelligent Force." In other words, true crop circles seem to somehow be related to UFOs and the extremely clever beings that control them.

Why? Because these circular patterns so closely resemble the impressions that are often left by flying saucers when they land in a crop field.

UFO FACTOID
Crop circles and saucer nests, similar in appearance, may actually be two different things.

In 1989, for instance, in Gulf Breeze, Florida, a large crop circle was found in a field of long grass near the sighting of a UFO. In another case, at Horseshoe Lagoon, Tully, Queensland, Australia, a farmer watched in surprise as a UFO flew out of his field (after stalling his tractor), leaving behind a "circular swirled imprint in the reeds" where it had originally landed.

This connection between crop circles and UFOs has made many people reject the phrase "crop circles" altogether. Instead, they prefer to call them "saucer nests," or "launch pads" for UFOs.

WITNESSES WATCH THE MAKING OF A CROP CIRCLE

On numerous occasions people have actually seen UFOs create circles in fields of grain as they stood and watched. Here's an example, taken from real casebook studies.

Casebook Study 35: On September 1, 1974, near Langenburg, Saskatchewan, Canada, at 11:00 A.M., a farmer named Edwin Fuhr was plowing his field when he noticed something strange in the distance.

As he got closer he was terrified to see five spinning domes hovering a foot off the ground. According to Fuhr, the metal objects looked like upside-down stainless steel bowls and were about eleven feet across. As he stared from his tractor in disbelief the UFOs quickly shot up into the sky and zoomed away. On the ground they had left behind five perfectly formed circles in his crop field.

> UFO FACTOID
> Crop circles, like UFOs, have been with us for thousands of years and show no sign of disappearing. Not surprisingly, just as UFO sightings are increasing each year, so are crop circles. While some are obviously hoaxes, many continue to defy rational earthly explanations.

The Royal Canadian Mounted Police investigated the event, interviewed Fuhr, and took photographs, pronouncing the case authentic.

In other instances other types of UFOs have been sighted and videoed around where crop circles were later found. Eyewitnesses, for example, have seen both NLs and DDs (basketball-sized balls of light) flying around, or even diving into, crop circles.

Attracted by their aura of mystery and beauty people travel from all over the world to see crop circles, even forming scientific research groups and large organizations of believers devoted to studying them. Some maintain that crop circles possess powerful spiritual energies that can heal, bring us closer to God or Goddess, or allow communication with the dead. Others think that they contain important messages from alien civilizations.

No satisfactory conclusion has yet been reached on where crop circles come from, what, if anything, they mean, or what their exact purpose is. But both cereologists and skeptics agree on one thing: like UFOs, we know that crop circles are real; we just don't always know who or what's behind them.

CHAPTER 7 AT A GLANCE

- Crop circles, or "saucer nests," are strange unsolved Earth mysteries, many which are directly connected to UFOs.
- Crop circles come in a wide variety of patterns, shapes, and sizes.
- Cereology is the scientific study of crop circles.
- Crop circles are often found near the sacred sites of ancient peoples.
- Crop circles are recorded as far back as the year 1590, but probably date from prehistoric times.
- There seem to be two basic kinds of crop circles: those that occur naturally (that is, they're self-generated), and those that are made by intelligent hands.
- Small brilliant spheres of light are sometimes seen darting around above crop circles.
- The beauty and artistry of crop circles fascinates many people.
- Crop circles may have the power to heal, and could be related to ancient Goddess-worship.
- Crop circles may be meaningless, but it's more likely that the authentic ones contain vital information, possibly from extraterrestrial beings.
- We don't know how crop circles are made, but theories range from the wind, elves, and fairies, to human hoaxers, the Devil, and UFOs.

8

BLACK SUITS, BLACK CARS: MEET THE MIB

"The matter [of UFOs and aliens] is the most highly classified subject in the United States Government, rating higher even than the H-bomb. Flying saucers exist. Their modus operandi is unknown but concentrated effort is being made by a small group headed by Dr. Vannevar Bush. The entire matter is considered by the United States authorities to be of tremendous significance." — Wilbert B. Smith, Canadian governmental radio engineer, 1950

CURIOUS & CURIOUSER: MEET THE MEN IN BLACK

While crop circles baffle and fascinate us there is an even weirder aspect of the UFO phenomenon: the men in black, or the **MIB**, as they're called for short. Just who, or should we say, what, are the MIB?

Their purpose, at least, seems to be to harass, scare, intimidate, and finally silence people who have seen UFOs, want to write about UFOs, want to research UFOs, or who just want to talk about UFOs.

Why would anyone want to silence people who're interested in UFOs?

Because, as we've learned, the governments of the world don't want anyone in the public sector to know that UFOs are real. They don't even want us talking about UFOs.

> UFO FACTOID
> Men in Black, or the MIB, are one of the more bizarre features of the UFO mystery.

The MIB use a number of different scare tactics to get their way, many of them as odd and scary as the MIB themselves.

UFO witnesses report that it was not uncommon to find MIB violently knocking on their front door on the same day of their sighting—even if they hadn't told a single person about it! Often the MIB will say that they're a member of **NORAD**, or some other government agency, and display fake-looking ID. They interrogate the person, demand and confiscate any photos, camera film, memory cards, or video of the UFO, and threaten her or him with prison—or even worse. Then they abruptly walk away, jump into a large black car, and speed off.

> UFO FACTOID
> The MIB often wear hats from the 1940s, drive old model black Cadillacs from the 1950s, and wear sunglasses from the 1960s. The reasoning behind this weirdness continues to baffle researchers.

Others say that the MIB come as black shadowy figures in the night, appear next to their beds, and paralyze them. Some UFO witnesses say that their cars have been forced off the road by MIB, who then warn them not to discuss their sightings with anyone.

There are things that are even weirder about the MIB, such as the way they dress, look, talk, and act. Here's a list of some of their more unearthly and bizarre characteristics.

- For some strange reason the MIB wear costly black suits, thin black ties, white shirts, and old-fashioned hats and shoes in the style of the 1940s and 1950s.
- They drive old-fashioned black Cadillacs or limousines that smell new, often with unissued license plates that can't be traced.
- Their skin is often unnaturally or eerily white.
- They usually appear to be either Asian or Scandinavian in appearance.
- They often travel in threes.
- They never blink their eyes and they move quite mechanically, like robots.

- Though some MIB are quite tall they're usually much shorter than the average person.

- They behave oddly and awkwardly, almost otherworldly, as if they're not from here.

UFO FACTOID
The Men in Black display numerous outlandish and even contradictory traits simultaneously. For example, clumsiness and confidence, politeness and disrespectful behavior. Most appear Asian, have pale skin, wear old-fashioned black suits, and travel in threes.

- The MIB usually speak very formally (often with a strange accent) and in a monotone voice, but sometimes they sound as if they're trying to imitate an old Humphrey Bogart movie from the 1940s.

- They ask unusual and even rude questions.

- The MIB often show amazement over everyday items such as eating utensils and ballpoint pens.

- They can appear or disappear at will.

- They use hypnotic techniques to get people to cooperate with them.

REAL LIFE ENCOUNTERS WITH MIB

The first known MIB report that's connected to UFOs comes from Wales.

Casebook Study 36: In March 1905 a number of unexplained lights in the sky were sighted by a group of startled Welshmen and women. Afterwards one of the women who had seen the lights was visited by a "man dressed in black." He came to her each evening for three nights in a row and delivered the same message each time.

Unfortunately we'll never know what that message was. The young woman was too scared to repeat it.

Since that time, more and more UFOs are being seen. In turn the MIB are appearing more frequently, more threateningly, and with more urgency. Here are a few more recent cases.

Casebook Study 37: For no apparent reason, in 1953, a Connecticut man, Albert K. Bender, suddenly closed down his popular UFO organization, the International Flying Saucer Bureau (or IFSB). When asked why, he would only say that "three men in black suits" had come to him, told him what UFOs are, and threatened him with prison if he ever told anyone the secret.

> UFO FACTOID
> The MIB can come across like hopelessly bad actors from a 1950s B-movie. Yet their presence often terrifies eyewitnesses.

Casebook Study 38: Not all MIB actually dress in black clothing. In a 1960s case a strange man wearing an Air Force uniform rushed a group of UFO witnesses (which included police officers) into a school room.

He then stared the surprised group in the eyes and said: "You did not see what you think you saw. So do not talk to a single person about your sighting." Then he quickly left as mysteriously as he'd arrived.

Casebook Study 39: On August 3, 1965, highway inspector Rex Heflin snapped four photographs of a UFO in Santa Ana, California. Later that day a mysterious man claiming to be from the "North American Air Defense Command" showed up at Heflin's home and demanded the film. Heflin never saw the man, or his photos, again.

> UFO FACTOID
> If you see a UFO you may receive a visit from the Men in Black—even if you never mention your sighting to a single person!

Casebook Study 40: In 1975 a Mexican pilot who had seen a UFO while flying his private plane, was on his way to do a TV interview about his experience. Suddenly he was forced off the road and menaced by four men in black suits who told him not to discuss his sighting. The pilot thought the incident was a prank and didn't give it another thought.

A month later he was on his way to talk with J. Allen Hynek, the famed American astronomer, when the MIB appeared again. This time the threat was far more serious. The pilot, realizing that this was no joke, swore he'd never talk about his UFO sighting with anyone again. And he hasn't.

SECRET AGENTS, ALIENS, OR DEMONS?: OTHER MIB RIDDLES

As to who or what the MIB are, no one knows for sure. Some think that they're simply intelligence agents working for a black government.

Others say that they're aliens trying to pretend that they're humans (and not doing a very good job of it).

Still others believe that the MIB are supernatural beings, something like demons, who can pass back and forth between the material plane and some other dimension.

Adding to the aura of mystery surrounding the MIB is the fact that the U.S. government says that these peculiar "agents" don't work for them—and never have. In an official 1967 memo from the Air Force Assistant Vice Chief of Staff Hewitt T. Wheless, the military claimed that it had no idea who they are. Instead it asked that anyone, military or civilian, who even hears of the MIB should immediately contact their local **OSI**.

> UFO FACTOID
> The MIB claim they work for the government, but the government claims they have no idea who they are. Just another one of the many riddles of the UFO phenomenon.

But how do we know this memo isn't just another part of the UFO Coverup? We don't.

One more enigma: why are there only men in black? Why aren't there any "women in black," or WIB? Like many other questions related to UFOs this too remains a problematic enigma.

CHAPTER 8 AT A GLANCE

- The MIB, or Men In Black, are one of the more bizarre aspects of the UFO phenomenon.
- MIB behave and look strange, wear clothes from the 1940s, and drive big black cars, like 1950s Cadillacs.
- MIB have occult powers which they use to control people.

- The purpose of the MIB seems to be to scare and silence people who've seen, or who want to openly talk about, UFOs.
- No one knows for sure who or what the MIB are, or where they come from. They may be secret government agents, or even aliens disguised as humans.
- If you even hear rumors about the MIB, let alone see them or come in contact with them, the U.S. government wants you to report it to your local OSI immediately.

9

UFOS ARE REAL: THE EVIDENCE

"More than 10,000 sightings [of UFOs] have been reported, the majority of which cannot be accounted for by any 'scientific' explanation, for example, that they are hallucinations, the effects of light refraction, meteors, wheels falling from aeroplanes, and the like. . . . They have been tracked on radar screens . . . and the observed speeds have been as great as 9,000 mph. I am convinced that these objects do exist and they are not manufactured by any nation on earth. I can therefore see no alternative to accepting the theory that they come from an extraterrestrial source." – Air Chief Marshall Lord Dowding, RAF, 1954

"ANGEL HAIR," EXPLODING LIGHT BULBS, & SCRAMBLED JETS: NINE PIECES OF EVIDENCE FOR UFOS

The Men in Black may often succeed in scaring witnesses into keeping quiet about their extraordinary experiences with extraterrestrials, but they can't hide the truth from the rest of the world: UFOs are real, as real as a Boeing 777, the space shuttle Discovery, the International Space Station, or an F-117 Nighthawk.

The plain fact is that literally millions of people from every walk of life have seen literally millions of objects in the sky that cannot be identified, and that behave in such a way that would be impossible for human-made aircraft.

Those who claim that UFOs are nothing more than products of vivid imaginations, or that they're mirages or optical illusions, or that they're just misidentified normal aircraft, are also the same individuals who complain that "there's no proof that UFOs are real."

In truth, we have an enormous amount of physical proof showing that UFOs are real and that they're visiting our planet.

Let's look now at the nine most important pieces of evidence.

UFO FACTOID
The U.S. government once tried to make a saucer-shaped aircraft. Known as the Avro-Car, or VZ-9V, it was certainly elegant looking. But in reality it was unstable, noisy, dangerous, and completely inoperable—quite unlike a real UFO. As a result, the entire project was soon scrapped. Despite our obvious and complete inability to construct even a rudimentary flying disk, skeptics continue to maintain that all saucer-shaped UFOs are secret military aircraft.

Proof 1: When UFOs land on the ground they almost always leave behind a "footprint" of some kind. These include: strange marks, swirled grass (crop circles), various imprints (usually of landing "feet" or pads), and burned areas (possibly from radiation) in the soil. Sometimes melted impressions are found on asphalt roads. If there are trees in the area the bark is often scorched, and branches and leaves are found to have been broken off.

Even if a UFO doesn't land it can still leave signs of its presence, such as "angel hair": fine white stringy threads that can look like cotton candy or Christmas tree tinsel. According to eyewitnesses this strange unknown substance, often found on the ground after a UFO passes overhead, mysteriously evaporates after a few hours.

On other occasions extremely small spheres are found on the ground that UFOs have flown over or landed on. Laboratory tests on these tiny pellets have established that they're of "unknown origin."

Humans simply don't make aircraft that leave behind these types of materials. And since they're always found around where UFO sightings occur it's obvious that these are real pieces of physical evidence of UFOs—all of which can be seen, smelled, touched, weighed, and measured.

Proof 2: When a UFO is present it almost always has a peculiar effect on things. Animals—such as cows and horses, or pets such as cats, birds, and dogs—become agitated and begin to act irrationally. As we saw in the Bentwaters-Woodbridge Incident, wild animals, such as deer and rabbits, have even been known to run frantically out of woods and fields, trying to escape the presence of UFOs.

People too can also be profoundly affected by UFOs. For instance, many UFO witnesses later report feeling tired, dizzy, or sick during or after their sighting, while still others say that they're energized by the experience.

Another bizarre fact is that UFOs nearly always cause electronic instruments to malfunction, or even stop working completely. This phenomenon, called the "electromagnetic effect," or **EME** for short, is very common.

For example, we have thousands of reports of car engines shutting off, radios becoming full of static, light bulbs popping, compasses spinning, and telephones going dead, when UFOs come into an area. And TVs have also been known to turn on and off by themselves when UFOs are nearby.

UFO FACTOID
If one day your car, TV, lights, computer, or phone shut off for no apparent reason, there may be a UFO in your area.

Not surprisingly, many of the world's major power failures have been linked to UFO activity, such as the Great Blackout of November 9, 1965, which hit the entire Northeast region of the U.S., leaving millions of people in total darkness.

Our space equipment seems to be effected by UFOs. In 1964 alone, for instance, four U.S. satellites suddenly stopping working. Several months later, all four of the broken crafts mysteriously came back online within a few days of one another. Perplexed space authorities guessed that they had been "knocked out" by meteorites. But if so, what, or who, repaired the damage?

Even time itself is influenced when UFOs are around. Many people report that after sighting a UFO they experience **missing time**, in which minutes, hours, or even days, pass that they can't account for.

There is no human-made aircraft that has these types of effects on electronic equipment, animals, plants, and people. And there never will be: even if we had the technology, what's the point of making something that no one would buy?

Proof 3: When UFOs are spotted near military installations jets are often "scrambled," or quickly sent off to chase or even shoot them down. Because a UFO is so fast no jet has ever been able to catch one, or even get close to one.

Missiles and rockets, as swift as they are, can't hit UFOs, and they've even been known to zig-zag to avoid bullets, many which travel at thousands of miles per hour. UFOs are so fast that when they're chased they'll often "wait" for the jet to catch up with them, then speed ahead again. Thousands of pilots have reported this exact experience, saying that UFOs seem to enjoy "playing tag" with them.

But even though they can't catch them, pilots do see the UFOs as they pursue them. They also film them with "gun-cameras" (cameras mounted on their wings) and track them on their aircrafts' radars. Furthermore, these same UFOs are often followed simultaneously by military and civilian personnel on ground-based radar.

UFO FACTOID
The military routinely scrambles jets to pursue suspected UFOs, often with the order to shoot them down. Why would our government follow this expensive, time-consuming, and dangerous policy if UFOs aren't real or if they're secret military projects?

Incredibly, the government possesses actual records showing that military jets that chase UFOs often crash, and that at other times the pilots and their aircraft simply disappear and are never seen or heard from again. In these type of situations there can no question that the pilots were chasing 100 percent certifiable UFOs.

In fact, the military doesn't bother sending up jets unless they're sure that it's a UFO. After all, why chase a known object, like a weather balloon?

SIGHTINGS BY PILOTS, CRASHED SAUCERS, & ANCIENT MYTHS

Proof 4: Military, private, and commercial pilots report many instances in which UFOs have paced or followed their planes and jets. Often the passengers see them as well.

These witnesses describe eerie metallic objects that zip alongside and then hover mysteriously in place just outside their windows. Some report seeing strange flickering lights inside the cockpits of these UFOs; other times, during nighttime sightings, the spooky craft will often sweep powerful searchlights across the entire length of the jetliner, much to the horror of the passengers and crew.

As part of their training, military and commercial pilots are taught to identify every known aircraft in existence. But they can't identify these strange aerial objects. This makes what they see outside their cockpits genuine UFOs.

On the other hand, if pilots are not seeing genuine UFOs, wouldn't they be fired from their jobs? After all, who wants to ride in a plane with a pilot who sees things that "don't exist"?

The truth is that pilots don't lose their jobs for reporting sightings of UFOs. Why? Because the military and the airline companies themselves know that these objects are real. And you can't legally fire someone for seeing something that's real. This is why pilots are actually only liable to lose their jobs if they discuss their sightings with the public.

UFO FACTOID
Commercial pilots report hundreds of UFO sightings each year. Typically their accounts are suppressed and they themselves aren't allowed to discuss the incidents. Failure to follow this order can result in job termination. Why all the fuss if "UFOs aren't real," as skeptics and the government tell us?

Proof 5: Many government employees are now coming forward and testifying that governments all over the world have recovered UFOs that have crashed on Earth.

According to these individuals at least forty crashed flying saucers have been captured by the world's governments so far. The remains of these craft have been locked away in secret airplane hangars where the military is studying them, and "reverse-" or "back-engineering" them, in an attempt to discover the secrets of their propulsion systems.

According to **JANAP** (a military legal publication), the penalty for revealing this information can bring fines of thousands of dollars, loss of one's job, and even long prison sentences without a trial. And yet these individuals still come forward.

Why would government employees risk so much if they weren't telling the truth? And why would the government impose such harsh penalties if there were no such things as UFOs to begin with?

Proof 6: Ancient myths describe meetings between people and divine beings from the sky, and prehistoric art depicts obvious space travelers and dazzlingly bright aerial craft landing on Earth, or shooting down beams of light.

As we've seen, the Bible itself is brimming with stories of close encounters with strange aerial objects and luminous beings. If people living today had these experiences we would call them "aliens" and "UFOs."

ABDUCTION MARKS, SECRET GOVERNMENT DOCUMENTS, & WORLDWIDE SIGHTINGS

Proof 7: Many people who've been abducted by aliens find that afterward their bodies have all kinds of new marks on them, from scoop-marks, holes, and scratches, to cuts, bruises, and burns. These marks aren't from visits to doctors or from accidents. So how do we explain them?

Additionally, some abductees find that their bodies have been implanted with some kind of strange tracking device, one that leaves no scar on the skin and creates no inflammation in the body. Doctors and scientists who remove and study the tiny objects are baffled by them, and can provide no logical explanation as to their makeup, function, or origin.

> UFO FACTOID
> Check your body for unusual markings, scars, and discoloration. You may be shocked to find a scoop-mark on your shin, a triangular scar on your arm, or a strange burn mark on your back. Ask your doctor what they are. You'll be surprised at his or her answer.

Since implants and fresh body marks only occur immediately after encounters with UFOs and aliens, they must be considered real physical evidence of an extraterrestrial presence here on Earth.

Proof 8: Thousands of pages of declassified governmental documents, obtained by American ufologists through the "Freedom of Information Act" (or FOIA), reveal a major UFO coverup: though the U.S. government publicly ignores, or even denies, the reality of UFOs and aliens, not only has it actually known about them since at least 1947, it's also extremely interested in them. Hundreds of similar documents from the United Kingdom show that the British government knew about UFOs even earlier, since 1943.

Proof 9: We've saved the strongest piece of evidence for last: millions of people have personally witnessed, photographed, filmed, and videoed millions of UFOs all around the world.

Many of these UFO sightings have not only been made by multiple witnesses—often by highly trained and experienced observers (such as people from the police, **MI6**, the **KGB**, the Army, the Navy, the Air Force, the **FBI**, and the **CIA**)—but they've also been seen on both airborne- and ground-based radar at the same time. Other professional witnesses include, as we've discussed, thousands of commercial and private pilots from all over the world.

UFO FACTOID
It's not easy to discount the thousands of eyewitness sightings of UFOs by levelheaded, highly trained individuals like police officers, pilots, and military officers.

In addition, people often say they can see sunlight glittering off the surface of UFOs. Other times, if it's over a lake or the ocean, the UFO's reflection is seen mirrored on the water.

Witnesses who've had real close encounters with UFOs often describe intricate designs on the outside of the craft, similar to the UFOs in the movie *Close Encounters of the Third Kind*. Still others report seeing the shadows of UFOs as they speed overhead or land on the ground.

These things tell us that what people are seeing aren't illusions. They're seeing real objects, objects that have solid mass, weight, and structure. In a moment—using actual UFO reports—we'll explore all nine of these pieces of evidence in greater detail.

UFO FACTOID
UFOs produce shadows on the ground, reflections on water, and tracings on radar, and are seen by multiple eyewitnesses, sometimes as many as a million people at once. Yet we continue to be told that "UFOs' don't exist."

FURTHER EVIDENCE FOR UFOS: MYSTERIOUS LIGHTS ON THE MOON

Nearly all of our evidence for UFOs comes from eyewitnesses who see UFOs as they're interacting with our planet: either on the ground, in our airspace, or in outer space around the Earth.

Few people realize, however, that our own Moon provides proof that UFOs exist. Some of this evidence dates back to a time long before we invented the electric light bulb, let alone spacecraft that could fly to the Moon.

UFO FACTOID
TLP are a common sight on or above the Moon. What else could they be if not UFOs? As of 2015 humans haven't established a base on the Moon.

Known as **TLP**, or "transient lunar phenomena," astronomers have long noted peculiar flashes of light, moving lights, mists and clouds, and strange colorful displays that appear and disappear on or above the Moon's surface.

This phenomenon is so obvious that by 1968 even NASA was forced to take notice. That year, after researching 570 strange but well-documented cases from between the years 1540 and 1967, NASA published its *Chronological Catalog of Reported Lunar Events* (also known as *Technical Report R-277*).

Examples of TLP, studied by NASA and others, include the following:

- As early as 1821 astronomers were reporting blinking lights in and around a Moon crater known as Aristarchus. The lights appeared again in 1825, and in 1835 the famous English astronomer Francis Bailey watched in amazement as a bright light moved around inside the crater. To put this in context: Bailey's sighting occurred sixty-eight years before the Wright brothers flew at Kitty Hawk, and 126 years before humans launched the first man, Russian cosmonaut Yuri Gagarin, into space in 1961.

- English astronomer William Frederick Denning (discoverer of five comets, several nebulae, and a new star in Cygnus) saw this same light twice in one year: on June 10, 1866, and again on May 7, 1867. And on January 23, 1880, noted French astronomer Etienne Leopold Trouvelot observed a thin thread of light stretched across Aristarchus.

UFO FACTOID
The last time humans stepped foot on the Moon was in December 1972, during the Apollo 17 Mission. Yet thirty-eight years later TLP are still being spotted on the Moon's surface. Who's up there—and why?

- On March 3, 1903, astronomers in both London, England, and Marseilles, France, were looking at the Aristarchus crater through their telescopes when all of a sudden a radiant light came on. It began to blink on and off, then finally disappeared.

Russian spectrograms, taken at the Pulkovo Observatory in 1958 and 1961—eleven years before humans (Apollo 11) landed on the Moon in 1969—revealed that gases such as hydrogen and carbon, were escaping from the Moon's surface near Aristarchus. Such gases indicate the presence of life. And yet conventional scientists refer to our lunar neighbor as being "dead."

Strangest of all was what astronomers from Lowell Observatory in Flagstaff, Arizona, saw on both October 29, 1963, and on November 27, 1963. As they trained their twenty-four-inch telescope on the Moon, they watched in disbelief as a number of bright red lights appeared north of the crater Herodotus. After disappearing, they then reappeared on the rim of Aristarchus. These two craters are several miles apart.

UFO FACTOID
The existence of TLP, as well as the presence of artificial structures on the Moon, leads some to believe that our lunar neighbor may be a great, hollow, artificially made "spacecraft," used by aliens as a base of operations.

As we'll see, American astronauts found even more convincing evidence of UFOs when they landed on the Moon for the first time six years later.

ARTIFICIAL STRUCTURES ON THE MOON

The Moon offers us other proof of the reality of UFOs and aliens as well. Careful researchers, such as Richard Hoagland, have discovered stunning photos of unknown artificial structures on the Moon, taken from NASA's own lunar atlas. The existence of these anomalies suggests the real possibility of nonhuman cities on the Moon. Let's look at just five examples.

"The Castle": During the Apollo 10 mission to the Moon the astronauts captured a huge complex object on film (frame 4822) that appears to be suspended several miles above the lunar surface by a cable. It's delicate construction leads some to believe it's "crystalline."

UFO FACTOID
NASA photo technicians carefully go over their photos before releasing them to the public, airbrushing out any unusual features on the Moon, such as the five anomalies listed here.

- "The Shard": Located near the Moon's Ukert region, and photographed by Orbiter 3, this bizarre structure is nearly 1.5 miles tall.

- "The Tower": Found in the Sinus Medii area and also photographed by Orbiter 3, this incredible feature is over five miles tall, and includes a strange cube-like object at its tip that's at least a mile wide.

- "The Triangle": Inside the Ukert crater lies something quite astounding: a perfect equilateral triangle, an object that forms the very foundation of tetrahedral geometry and hyperdimensional physics. Large *perfect* equilateral triangles are not found in Nature.

- "The Bridge": In 1953 science writer John O'Neill discovered what appears to be a bridge twelve miles long spanning the Mare Crisium crater. Naturally space scientists ridiculed O'Neill; that is, until his finding was authenticated by British astronomer Dr. H. P. Wilkens just a few weeks later.

None of these highly sophisticated and intricate structures can be explained away as "natural," so obviously they must be made by a superior intelligence. But whose?

Humans have not spent enough time on the Moon to have built even a small military base let alone enormous "towers," "castles," and "bridges" on and above its surface.

UFO FACTOID
The hundreds of cases of mass sightings of genuine UFOs by literally millions of people are impossible to ignore. Yet skeptics routinely disregard them.

AUTHENTIC MASS UFO SIGHTINGS

While skeptics have a hard time refuting moonlights, satellite photos, and military UFO sightings, they find it even more difficult to dismiss another type of sighting: the **mass sighting**.

Skeptics like to point out that UFOs never appear over heavily populated areas, or large metropolitan cities. But in fact they do. And more often than you might think. Here are a few casebook studies, taken from genuine UFO files.

Casebook Study 41: On December 15, 1980, thousands of people watched a triangular UFO with an orange nose, a silver body, and a blue rear section, hover over downtown London, England. As the crowds stared in amazement the brilliant object divided into two sections. These moved around each other for a few moments, then shot straight up into the sky and disappeared, leaving two very distinct vapor trails.

UFO FACTOID
Many people believe that the Vatican UFO Incident was religious in nature, owing to the large cross that was formed by the strange objects.

Casebook Study 42: On September 5, 1968, thousands of people crowded the streets to look at a super bright triangular-shaped UFO that was soaring above the city of Madrid, Spain. Roads leading in and out of the city were backed up for miles with traffic jams as alarmed onlookers stopped to observe the blinding object.

Casebook Study 43: In 1967 thousands of people watched a brilliant globe-shaped UFO hurl across the sky over the city of Peking, China, at speeds far beyond those of human-made aircraft. As the huge crowds stared in disbelief, the bizarre object paused, hovered, and then zoomed off into the night sky.

Casebook Study 44: On June 14, 1980, a strange UFO appeared over the middle of the city of Moscow, Russia, where it was witnessed by crowds of thousands. The crescent-shaped craft was described as huge and orange-red in color. As it flew it gave off swirling clouds of glowing gases.

Casebook Study 45: At noon on November 6, 1954, thousands of people stood in awe as squadrons of UFOs that looked like "white dots," flew in complicated well-kept echelon formations over the city of Rome,

Italy. Coming in from opposite directions, two V-shaped squadrons (of fifty UFOs each) met directly over the Vatican.

As they came together, the two groups formed themselves into a perfect cross, which then turned slowly like a giant Ferris wheel on its side. The squadrons then broke apart and sped off in opposite directions.

A few minutes later the *dischi volanti* (Italian for "flying saucers") returned and repeated the entire performance. This time long glassy strands of "angel hair" fell from the sky. Baffled witnesses reported that the strange material completely "evaporated" a few hours later.

Casebook Study 46: Since the 1990s hundreds of UFOs have been seen, photographed, and videoed flying, diving, sailing, and hovering over one of the world's most heavily populated metropolises: Mexico City, Mexico (population: 9,000,000). As a result, some of these sightings are witnessed by literally millions of people at the same time. So many UFOs have been seen by so many people in "The City of Palaces" that local schools now offer classes on ufology due to the enormous interest that's been generated.

Not all mass sightings are made by adults. Many have been made by children and young adults, as the following example shows.

Casebook Study 47: In 1995, in Zimbabwe, South Africa, all of the children at the Ariel School saw a brilliant oval-shaped craft land at the far end of their playground. As the children watched, two creatures with large heads and small bodies emerged from the glowing object. According to the children's testimonies, when the alien beings moved around it looked like they were running in slow motion, as if they were on the Moon.

This case is particularly fascinating because in this part of the world the public had almost no TV or magazine exposure to the topic of UFOs and aliens up to 1995. And yet all sixty-

> UFO FACTOID
> Because the eyewitnesses were children (each who gave the same description independently), the Zimbabwe Incident is one of the most convincing of all UFO/alien encounters.

two children described the same type of spacecraft and strange creatures that are being seen in other more developed countries. Since they didn't know what the object was or what the creatures were, they reported exactly what they saw. This makes these children very believable eyewitnesses.

Not all mass sightings are made by civilians. Some, as we've seen in the Bentwaters-Woodbridge Incident, are made by the military. Here's another example.

Casebook Study 48: On the night of October 28, 1978, hundreds of fighter pilots were outside watching a movie on their airfield at Lintiao Air Base in the Gansu Province of China.

In the middle of the film something in the sky caught their eyes. Looking up they stared in disbelief as an enormous glowing UFO soared slowly just above their heads. It was quite long, outlandish in appearance, and had two brilliant searchlights on the front end.

A few minutes later it lazily disappeared over the horizon. The pilots, who are trained to identify every type of aircraft, said that they'd never seen anything like it before.

UFO FACTOID
Non-believers say the Lintiao Air Base UFO was a "secret "experimental military aircraft." But if so, why would the military fly it over its own troops?

Not all mass UFO sightings are modern. The following famous case dates back almost 500 years.

Casebook Study 49: On August 7, 1566, a huge fireball and dozens of strange black globes soared over the city of Basel, Switzerland, filling the townsfolk with terror. Thousands watched as the objects hovered and streaked mysteriously through the sky. Articles and drawings of the sighting were published in the local paper, the *Basel Broadsheet*.

Despite these amazing reports—which can only be described as incredible sightings made by a credible number of people—skeptics say that these millions of UFO eyewitnesses don't represent evidence that UFOs really exist. Why?

Skeptics don't consider eyewitnesses of UFOs to be reliable. Instead, they disparagingly refer to such reports as "soft evidence," claiming that what is needed to prove the reality of UFOs is "hard evidence"; that is, a crashed saucer or a dead alien.

> UFO FACTOID
> When it comes to UFOs skeptics reject eyewitness testimony, the same kind of testimony that can determine the life or death of criminal defendants.

But the testimony of any one of these eyewitnesses would stand up in a court of law and could easily get a criminal put to death. So why don't they count in the "court of science"?

"Because," a mainstream scientist might counter, "the standards of proof required by science are more rigorous than those required by law." However, as even the most unbelieving scientist should know, *the absence of evidence is not necessarily evidence of absence*—as we're about to see.

SCIENTISTS & GOVERNMENT OFFICIALS WHO BELIEVE IN UFOS
Not all people with backgrounds in the military or the government are skeptics, nonbelievers, and debunkers. Far from it.

In fact, many—like President Ronald Reagan and President Jimmy Carter, the latter who's also a nuclear physicist—not only believe in UFOs, they've also seen UFOs. (And they weren't alone at the time: in both incidents these were multiple sightings.) Here's a sampling of what some of these important high level individuals have to say on the subject.

> UFO FACTOID
> The private correspondence of President Harry Truman, the founder of the MJ-12, reveals that he was obsessed with the topic of UFOs, and that he believed they're real and pose a potentially serious security risk to the U.S.

Nick Pope (the "real Fox Mulder"), former Senior Executive Officer at Britain's **MOD**, and the author of the foreword of this book, said that he entered his job a UFO skeptic but left it a believer. Why? Because of the overwhelming evidence, much of it garnered from highly educated and highly trained individuals.

In 1962 U.S. Admiral Roscoe Hillenkoetter, the first director of the CIA, told the public that there is no country on Earth that has produced aircraft that can match the speeds and maneuverability of UFOs. Behind closed doors high ranking military officers are deeply worried about UFOs, he said, but citizens are persuaded that they don't exist due to "secrecy and ridicule."

Stanton T. Friedman, a nuclear physicist who's worked at General Electric, Westinghouse, and McDonnell Douglas, states without reservation that "UFOs are real!" Friedman, known as the "Father of Roswell," is on record saying that he is fed up with the Air Force's obfuscation and that it should stop lying to the American public, the media, and Congress about UFOs.

UFO FACTOID
Nuclear scientist Stanton T. Friedman is willing to debate the Air Force on the topic of UFOs "anytime and anyplace."

Albert M. Chop, deputy public relations director of the National Aeronautics and Space Administration in 1965, said he's convinced that UFOs are not from our planet and that we're being carefully monitored by aliens.

Dr. Maurice A. Blot, famed 1950s mathematical physicist, states that the most rational explanation is that UFOs are real and that they're controlled by beings from somewhere outside our Earth.

U.S. President Gerald Ford believed in UFOs. At one point, fed up with the Air Force's lackadaisical attitude toward unidentified flying objects, he suggested that the government set up a special committee that did nothing else but investigate UFOs.

U.S. Senator Barry Goldwater, former Brigadier General of the U.S. Air Force Reserves, said that he believed UFOs to be authentic extraterrestrial craft.

Dr. Herman Oberth, famed German rocket scientist, maintained that UFOs could only have been constructed by creatures far more intelligent than humans, and that these being probably came from somewhere outside our solar system, or even outside our galaxy.

USAF Colonel Howard Strand (who had three encounters with UFOs in his F-94 fighter jet), said that he felt that UFOs were here gathering intelligence and that the beings operating them were an "advanced civilization."

Wernher von Braun, the German-American rocket scientist who is known as "the Father of Modern Rocketry," maintained that in the future the world will find it impossible to deny the reality of UFOs.

Dr. Carl Jung, the celebrated Swiss psychiatrist, maintained that UFOs are real, physical objects that behave outside the laws of physics, and which can only be operated by a scientific technology far beyond our own. Troubled by the worldwide governmental secrecy involving UFOs, he argued that the coverup had more of a negative impact on people than knowing the truth.

UFO FACTOID
Ignoring the seriousness of UFOs means disregarding the beliefs, knowledge, opinions, wisdom, and experiences of some of the world's greatest minds.

George Langelaan, member of the French Secret Service in 1965, was adamant that UFOs are genuine and that they come from outer space.

Dr. Charles Harvard Gibbs-Smith, aeronautical historian at the Victoria and Albert Museum in London, England, held that there are other races far older and more intelligent than our own who have been visiting us in what we call "UFOs" for a very long time.

Air Chief Marshal Lord Hugh Dowding, commanding officer of the Royal Air Force of Great Britain during World War II, was of the same belief. UFOs can only be from an "extraterrestrial source," he asserted in 1954.

In 1990 Mikhail Gorbachev, former leader of the USSR, said publically that UFOs are real and that they should be approached in a serious scientific manner.

Dr. James McDonald, senior physicist at the Institute for Atmospheric Physics, said that while he had no idea where UFOs originate he knew that it was not from anywhere on Earth.

Admiral Gerson de Macedo Soares, Navy general secretary of Brazil, says that he personally believes in UFOs, no matter what anyone else thinks.

Admiral Delmer S. Fahrney, head of the U.S. Navy guided-missile program in the 1950s, felt that, based on trustworthy reports, UFOs are authentic and are being constructed and operated by highly advanced beings.

UFO FACTOID
During the Apollo 15 Moon Mission a disk was clearly seen hovering over the Moon's surface.

Dr. Walter Riedel, another rocket scientist of note and a lead designer of Germany's famed V-2 rocket, believed that UFOs have "an out-of-world basis."

Lord Hill-Norton, Britain's former Chief of Defense Staff, said the "U" in UFO should stand for "unexplained," not "unidentified."

We'll recall that General Douglas MacArthur, famed general of the entire U.S. Army during World War II, believed that the next major world conflict would be an "interplanetary war," one fought between humans and aliens.

Lieutenant Colonel Philip J. Corso, a well-respected artillery commander, spy hunter, adviser to President Dwight Eisenhower, and intelligence officer who worked at the Pentagon, says that he personally saw dead aliens in crates being stored at Fort Riley (Kansas), and that Eisenhower later assigned him the job of reverse-engineering alien technology recovered at Roswell. Corso's book on his experiences with UFOs and ETs, entitled *The Day After Roswell: A Former Pentagon Official Reveals the U.S. Government's Shocking UFO Cover-up*, became a *New York Times* bestseller.

In 2010 one of the world's leading physicists, cosmologists, and space scientists, Stephen Hawking, went public with his belief that intelligent alien beings exist outside our planet. We should avoid contact with them, however, Hawking warned, as they will most likely be hostile.

In order to accept conventional science's view that there really are no such things as UFOs or aliens, we would also have to reject the words of these distinguished world leaders and scientists.

ASTRONAUTS & UFOS

There is one more piece of convincing evidence for UFOs: many astronauts—the most highly trained and respected of all aircraft observers—have witnessed UFOs, not only on Earth, but also in space.

> UFO FACTOID
> Astronauts are the world's most astute observers of air- and spacecraft. Yet skeptics completely dismiss their UFO sightings. This is not good science.

Casebook Study 50: One astronaut, Mercury 7's Colonel L. Gordon Cooper, for example, says that he actually became an astronaut because of his interest in UFOs.

While he was a fighter pilot in Germany in 1951 Cooper not only saw UFOs, he chased them. Later, in a 1978 letter to the United Nations, he described the experience. "We spotted UFOs for several days in a row, he noted, metallic saucers that flew at high altitudes and at extremely high speeds, maneuvering beyond the capabilities of our fighter jets." Cooper went on to write that he firmly believes in the authenticity of UFOs and that they derive from a superior species.

Casebook Study 51: During James Lovell's flight on Gemini 7 he radioed the ground with this message: "**Bogey** [UFO] at 10 o'clock high . . . We have several . . . actual sighting."

Casebook Study 52: Astronaut Scott Carpenter took a photo of what appears to be a strange unidentified flying object on May 24, 1962, during his flight aboard Mercury 7.

Casebook Study 53: In 1951, while test-flying a P-51 fighter jet in Minneapolis, Minnesota, Mercury astronaut Donald Slayton saw what he thought was a kite or a weather balloon up ahead. Traveling at 300 mph Slayton quickly caught up with the object, but found that it wasn't either. It was a tiny disk-shaped UFO, a mere three feet in diameter. As he approached the strange craft it went into a steep climb, accelerated, and disappeared above him.

UFO FACTOID
During the Apollo 16 Moon Mission many UFO-like objects were seen, such as a saucer-shaped craft filmed during lunar approach.

Casebook Study 54: Astronaut James McDivitt says that while he was in space he had "seen a lot of objects that I could not identify." One of these appeared outside the front window of his spacecraft on June 4, 1965, as he was soaring above the Earth on the Gemini 4 flight. McDivitt described the strange craft as cylindrical with a long pole-like object extending from one end. When the astronaut turned away and looked back, the UFO was gone.

Casebook Study 55: While aboard Mercury 8, astronaut Walter Schirra saw a number of UFOs pacing their space capsule, objects that NASA secretly code-named "Santa Claus." Later, on the 1968 Apollo 8 mission, James Lovell abruptly announced: "Please be informed that there *is* a Santa Claus." At the time only the astronauts and NASA scientists knew what he was really talking about.

Casebook Study 56: During their record-breaking landing on the Moon in July 1969 it's reported that Apollo 11 astronauts Neil Armstrong and Edwin "Buzz" Aldrin saw bizarre lights in the distance. On closer inspection they realized that two giant sinister-looking UFOs were sitting on the edge of a Moon crater looking back at them. According to some sources, after informing Mission Control (MC) about the threatening craft, MC radioed back:

> MC: "What's there? Mission Control calling Apollo 11."
>
> Apollo 11: "These babies are huge sir . . . enormous. . . . Oh God, you wouldn't believe it! I'm telling you there are other spacecraft out there . . . lined up on the far side of the crater edge . . . they're on the moon watching us. . . ."

After the astronauts took photos and film of the UFOs from both inside their capsule and outside on the surface of the Moon, the two objects lifted up off the crater rim and disappeared.

> UFO FACTOID
> NASA uses many different code words for UFOs, such as "bogey" and "Santa Claus," in order to hide astronaut sightings from the public.

One former NASA official, Maurice Chatelain, says that many of NASA's space flights, like the Gemini and the Apollo missions, were *always* tailed by UFOs, but the astronauts were not allowed to tell anyone. According to Chatelain, UFOs usually maintained a large distance between themselves and the astronauts' capsule, but other times they came very near, close enough to be easily seen with the naked eye.

Casebook Study 57: For instance, on two of his missions into space Dr. Story Musgrave (the only astronaut to have flown on all five Space Shuttles), says he saw what he believes may have been a living being. The alien creature, or "snake," was between six and eight feet long, flexible, and followed the astronauts for "a rather long period of time."

Casebook Study 58: Non-astronauts who fly research aircraft have also had UFO sightings. In April 1962, during NASA pilot Joseph A. Walker's test flight of the famous rocket-propelled X-15 aircraft, he saw and filmed six disk-shaped objects, none of which were ever explained.

UFO FACTOID
As the astronauts aboard Discovery STS-51-A made their approach to the WESTVAR VI satellite, the space shuttle's video camera captured an unknown gray sphere speeding past from left to right. The bizarre object appeared to be monitoring the shuttle's movements.

Casebook Study 59: A few months later, on July 17, 1962, Major Robert White also flew an X-15. When he reached a height of fifty-eight miles he looked out and saw a gray UFO pacing him only thirty feet away from his aircraft. The startled pilot is alleged to have radioed to the ground: "There are things out there. There absolutely is!"

UFO FACTOID
On September 15, 1991, Discovery's video camera recorded a number of distant luminescent objects performing maneuvers that reveal intelligent control. NASA claims that these objects are "ice crystals," an impossibility. Open-minded (true) scientists later calculated that one of the objects moved from a dead stop to 2,500 mph in one second! Revealingly, shortly after this controversial tape was made public, NASA discontinued live space transmissions.

Casebook Study 60: American astronauts aren't the only ones who've seen UFOs. On October 13, 1964, the Russian crew of the Voskhod 1 returned to Earth with harrowing tales of strange UFOs in outer space.

According to the three cosmonauts, Vladmir M. Komarov, Konstantin Petrovich Feokistov, and Boris Yegorov, they were continuously paced by incredibly fast "flying disks" that hit their space capsule with violent magnetic pulses of some kind.

Echoing Russia's former leader Gorbachev, who believes UFOs are genuine, another Russian cosmonaut, E. V. Khrunov, said that UFOs are definitely real—and extremely dangerous.

Apollo 14 astronaut Captain Edgar D. Mitchell sums up the opinion of many astronauts and pilots this way: "Everyone knows that UFOs exist. The only question now is where do they originate?"

THE VIDEO & PHOTOGRAPHIC EVIDENCE

UFOs seem to be spotted on *all* space missions, though this fact is carefully kept from the public. UFOs were sighted, and video and/or still photos were taken of them, on the following (highly abbreviated and alphabetized) list of space flights:

Apollo 8
Apollo 10
Apollo 11
Apollo 12
Apollo 13
Apollo 14
Apollo 15
Apollo 16
Apollo 17
Gemini 4
Gemini 11
ISS (International Space Station)
Lunar Orbiter IV
Lunar Orbiter V
Mercury 8
SkyLab III
STS-28
STS-29
STS-48
STS-51
STS-61
STS-64
STS-73
STS-75
STS-80
STS-88
STS-96
STS-100
STS-108
STS-109
STS-110
STS-114
STS-115
STS-116
STS-119
STS-123

The following list is of actual video and photo frames from some of the above mentioned missions. The key to the entries is mission/roll/frame, or mission/magazine/frame. For example:

STS-88-724-65 means: *Space Shuttle Mission 88, roll 724, frame 65.*
AS07-07-1835 means: *Apollo Mission 7, magazine 7, frame 1835.*

(For more information on these images, visit: http://eol.jsc.nasa.gov and also www.lpi.usra.edu)

AS07-07-1835
AS08-13-2319
AS08-13-2344
AS08-16-2593
AS09-21-3212
AS10-28-3988
AS10-28-3989
AS10-28-3990
AS11-36-5319
AS12-53-7927
AS13-61-8865
AS14-66-9290
AS14-66-9295
AS14-66-9301
AS17-134-20424
AS17-136-20853
ISS-01-376-004
ISS–006-E-39581
ISS-008-E14288
ISS-013-E-4000
LO2-M-162
STS-28-72-40
STS-61C-31-002
STS-61C-31-003
STS-61C-37-47
STS-88-724-65
STS-88-724-66
STS-88-724-67
STS-88-724-68
STS-88-724-69
STS-88-724-70
STS-96-706-1
STS-100-E-5015
STS-100-E-5220
STS-108-E-5380
STS-109-315-016
STS-110-E-5912
STS-115-E-07201
STS-116-E-05364
STS-115-E-07201

CHAPTER 9 AT A GLANCE

- We know that UFOs are real because they've been seen by literally millions of people.
- We have lots of other evidence that UFOs are visiting Earth, from the "footprints" they leave on the ground, to the strange marks people find on their bodies after seeing UFOs.
- UFOs can also affect electronic equipment, people, animals, and plants.
- UFOs are not only seen by the naked eye, they're also photographed, videoed, and detected on radar.
- Many people see the shadows of UFOs on the ground, or the reflection of UFOs on water. Others see the Sun shining off the sides of UFOs. This means that UFOs are solid objects.
- Countless numbers of mass sightings have occurred all over the world in which a UFO has been seen by thousands, or even millions, of people at the same time. Many mass UFO sightings occur over military installations and airports.
- Not all government and military people disbelieve in UFOs. Many government officials and high level workers not only completely accept the reality of UFOs, they've also seen them. They don't usually report their sightings, however, for fear of ridicule and losing their job.
- Many astronauts—the world's best observers of aircraft—have seen, filmed, and photographed UFOs, both on Earth and in space.
- Many of these videos and photos are now available to the public, hard evidence of the UFO phenomenon for all to see and study.

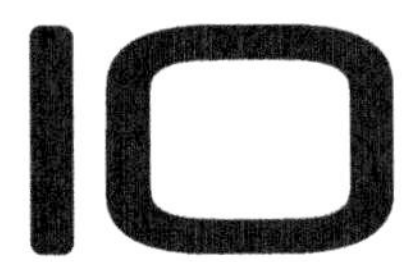

SKEPTICS, DEBUNKERS, & FALSE SCIENCE

"No agency in this country or Russia is able to duplicate at this time the speeds and accelerations which radars and observers indicate these flying objects are able to achieve. . . . there are objects coming into our atmosphere at very high speeds . . . [using] a tremendous amount of technology of which we have no knowledge." — Admiral Delmer S. Fahrney, former head of the Navy's guided-missile program, 1957

WHY SOME PEOPLE DON'T BELIEVE IN UFOS

Despite overwhelming evidence—as detailed in the previous chapters—showing that UFOs are real solid objects that originate from outside our planet, some people still refuse to accept their reality, a reality called **ETH** (the "extraterrestrial hypothesis"). Part of the problem is UFOs themselves.

> UFO FACTOID
> Are UFOs real or fake? Most scientists automatically assume that all UFOs are hoaxes, hallucinations, or secret government projects, without ever inspecting the evidence.

Genuine UFOs are so peculiar, so incredible, and so unlike anything in our everyday human experience, that they just don't seem possible. After all, if you really think about it the idea of alien spacecraft visiting Earth can seem "quite illogical," as *Star Trek's* Mr. Spock might say.

Ufologists refer to the general weirdness that surrounds real UFOs as "high strangeness." And it's partly because of this very strangeness that most scientists refuse to believe in them. Such people are called "nonbelievers," or "**skeptics**," because they don't accept anything as "real" unless it meets up to standards they deem "scientific"

UFO FACTOID
Even though there are still many unanswered questions about Jupiter's Great Red Spot (a weather system larger than Earth), scientists believe it's real because they've photographed it. Thousands of UFOs have also been photographed, and yet most mainstream scientists don't believe in them, and won't even examine the photos. Why? The answer is much simpler than you might imagine.

(these standards often differ from skeptic to skeptic, and from scientist to scientist).

Some skeptics are so threatened by the idea that UFOs are real that they'll go to almost any length to prove that they're not. These people, called debunkers, are traditionally quite close-minded and usually refuse to even look at the evidence for UFOs. Even when, on rare occasions, they do, they will usually do so with many preconceived notions, changing and twisting the evidence to fit their own ideas of how things should be.

Debunkers and other types of skeptics feel that everything in the Universe operates under certain basic natural laws, such as "the law of gravity," for example. UFOs, however, not only defy the law of gravity, but many other physical laws as well, which is the main reason skeptics don't accept UFOs as being authentic.

Instead, they maintain that most UFO sightings are actually secret experimental military aircraft, and that "encounters with aliens" are just "brain-generated" experiences. In other words, they think that people only see UFOs and aliens because they want to, not because they're actually seeing them.

And yet most UFO witnesses say they didn't believe in UFOs until they saw one. Some eyewitnesses even state that they never wanted to believe in them. This makes this particular claim by skeptics quite ridiculous.

This skeptical idea, which we'll call the "believing is seeing" theory, states that nearly all UFO sightings can be explained as one of the three "h's": a hoax, a hallucination, or hysteria.

UFO FACTOID
During his long and distinguished career U.S. Major Donald Keyhoe, head of the now defunct NICAP and author of the book *The Flying Saucers Are Real*, fought courageously against scientific skepticism.

UFO FACTOID
Even though we have abundant proof that UFOs have been visiting Earth for thousands of years, mainstream scientists believe that we must send probes, like Voyager, into outer space in order to search for extraterrestrial life.

HUMANS DON'T HAVE THE TECHNOLOGY TO BUILD & FLY UFOS

But the nonbelievers, the debunkers, and the skeptics are wrong. UFOs are as real as you and me; they're as real as your house, your car, your TV, or your pet dog or cat. We know that UFOs are real because people have seen them up close, photographed them, videoed them, and touched them. Thousands of people even say they've ridden in them.

Certainly no one who's seen a UFO will ever forget it, nor will they ever doubt the reality of UFOs again. They become what are called "**believers**."

I'm a believer myself. Why? Because I've seen many UFOs in the skies and there's no doubt in my mind that humans could not have created them. Here's an example.

On August 2, 2005, I needed to travel out-of-town for business and found myself on a Boeing 737. As we flew over West Virginia I was looking out of my window and spotted an object traveling at a blazing speed through the sky at approximately ten miles distant. At first I assumed it was a jet of some kind, but then I noticed that it's vapor trail was white while all of the other jets I had seen left black or gray vapor trails as they sped past us.

But something much more spectacular caught me eye: the object moved in a flat (straight) line at lightning speed across the entire visible horizon of my vision, approximately fifty miles, in about five seconds. This means it was traveling at a blazing 36,000 mph! We simply don't have the technology to move at anywhere near this rate of speed in Earth's atmosphere (because of friction); only in outer space where there's no air.

UFO FACTOID
An important ancient European drawing, found in Uzbekistan, plainly shows some kind of glowing spacecraft coming down from the sky. Skeptics, however, tell us that it's just "meaningless symbolism."

The mysterious object was level with our jetliner, which was flying at a mere 530 mph, at an altitude of about 30,000 feet (just under six miles up). This is well below the level of outer space, which begins at around one-hundred kilometers, or sixty-two miles, up.

Due to its appearance, speed, altitude, and behavior, this couldn't have been a meteor, weather balloon, satellite, flock of birds, swamp gas, ball lightning, secret experimental military aircraft, Saint Elmo's Fire, or my imagination (the person sitting in the seat next to me also saw it). What then was it? A UFO, plain and simple.

Of course people have seen far stranger objects in our skies. Do we have the knowledge, for example, to make huge spacecraft that can fly at thousands of miles an hour in complete silence, make ninety-degree turns without slowing down, and disappear and reappear in the twinkling of an eye? You can be sure that we don't.

Oddly, skeptics continue to claim that these kinds of aerial vehicles are made by people. But any country that had this amazingly advanced technology could, and would, easily rule the world. Is there currently such a nation? There isn't.

And even if such a country did exist, not only would they already be governing our entire planet both militarily and economically, but our skies would be full of exotic silent aircraft that could travel at tens of thousands of miles an hour in Earth's atmosphere and become invisible with the press of a button.

All of this is currently impossible at our present level of scientific knowledge, of course, and probably will be for hundreds, if not thousands, of years into the future. We may never fully master some of these feats.

But the builders of UFOs have, and they did it long before humans built the steam engine, let alone learned to fly on the sandy beaches of Kitty Hawk, North Carolina.

UFO FACTOID

Some 20,000 new species of animals are discovered each year, and we still don't know exactly why cats purr, what dolphins are saying, or how lightning is formed. We've only explored less than 1 percent of the world's oceans. In fact, we know more about the Moon than we do about the Pacific Ocean. Clearly, the Universe still holds many secrets. Despite this, mainstream scientists seem to think they already know nearly everything there is to know. This type of arrogance has never been part of *authentic* science, and never will.

HOAXES, HALLUCINATIONS, & HYSTERIA: DEBUNKING THE DEBUNKERS

In truth, the explanations given by skeptics for UFOs and aliens just don't hold up.

For one thing, the USAF says that it doesn't fly experimental aircraft over populated areas, and never has. And for good reason. Not only would the Air Force be endangering innocent civilian lives, but once the public saw these craft the military would no longer have the advantage of secrecy.

UFO FACTOID
Scientists once believed that the Earth is the center of the Universe. In reality, our tiny planet is located in an outer area of an outer arm in the outer regions of the Milky Way galaxy, far from the center of anything. Scientists of the future will also discover that their skeptical forerunners were wrong about UFOs.

As far as UFOs go, not only do many of them not do anything to conceal themselves, they actually go out of their way to get attention. Some UFOs even flash their lights at observers on the ground or pursue them in cars and planes.

Lastly, large stealthy UFOs were being seen long before the human invention of flight.

Obviously then, most UFOs are *not* secret human-made aircraft, military or commercial.

As for the skeptics' three "h's", let's now look at these more closely.

1. The Hoax Theory: First, it's true that some UFOs have been the products of jokes, deception, and general trickery—usually perpetrated by skeptics themselves. But these are in the tiny minority. Actually, only 1 percent of the UFOs reported to the military turn out to be hoaxes. And after all, not even the smartest pranksters could hoax the kinds of things that most UFO witnesses see. We humans simply don't have this kind of knowledge or technology.

UFO FACTOID
Scientists give many reasons for why they think UFOs are not real. What they don't seem to realize is that their reasons are so flawed, and often so blatantly absurd, that they actually only make themselves look ridiculous and the UFO hypothesis more believable.

2. The Hallucination Theory: Second, we know for sure that true UFOs aren't hallucinations because so many of them are seen by more than one person. As psychology has shown, it's

impossible for two people to imagine that they're seeing the same thing at the same time. More problematic for skeptics in this regard is the fact that in nearly every multiple sighting the individuals involved all describe or draw the exact same object, even though they've never met and live many miles apart.

3. The Hysteria Theory: Third, we can discount the skeptics' explanation of hysteria (meaning someone possessing extreme suggestibility and irrational behavior) because the vast majority of people that see UFOs are serious, trustworthy, sincere, healthy-minded individuals who have no desire for attention, fame, or money. These would include police officers, airline pilots, school teachers, jet fighter pilots, judges, office workers, doctors, ship captains, construction workers, truck drivers, and astronauts. Military personnel can actually be put in prison for discussing their UFO sightings with the public, yet many still come forward. Why would they risk so much for something if it wasn't true?

Skeptics claim that there's another type of hysteria at work behind the sightings of UFOs: mass hysteria, in which a group of people experience hysteria at the same time or during the same general period of time.

> UFO FACTOID
> Black holes violate the known laws of physics, but they're now considered real by nearly all space scientists. Still, black holes are far from being fully understood. Why can't mainstream scientists have the same attitude toward UFOs, which also defy currently known laws? The answer is, in part, that they approach UFOs emotionally and personally instead of intellectually and dispassionately, the way they approach black holes.

The problem with this explanation is simple: not only are most UFOs seen by one person, but mass hysteria is usually a short-term phenomenon, lasting anywhere from a few hours to, in extremely rare cases, a few years. And yet the modern UFO wave has been going on for over a half a century.

Additionally, UFOs were being sighted by ancient, and even prehistoric, people many thousands of years ago. Clearly then UFOs are not a modern-day phenomenon, nor are they a product of mass hysteria.

In truth, we can safely say that most UFO witnesses are not hysterical, they're not seeing or creating hoaxes, and they're not having hallucinations. And because UFOs predate the modern world and far outperform our current aircraft technology, we can also state that most UFOs are not secret or experimental military craft.

UFO FACTOID
We've found out recently that parallel universes are real, and that there are an infinite number of them, each one representing the different possible outcome of every possible action. This was an idea scoffed at by most scientists only a few years ago.

Adding to this is the fact that many UFOs are also picked up by military and commercial radar: a radar sighting is not a hallucination, nor can it be hoaxed.

And what about those unexplained flying objects that are seen by people on the ground, pilots in the sky, radars on the ground, and radars on airborne jets, *all at the same time*? These cannot be dismissed with a wave of the hand. Such sightings demand serious scientific attention.

Yes, it's true that there are many mysteries surrounding UFOs: we still don't know exactly who makes UFOs, who flies them, what propels them, what they're made out of, why they're here, or where they come from. But despite these unanswered questions we do know what UFOs aren't: they aren't imaginary.

UFOS AND THE OPEN-ENDED SEARCH FOR TRUTH

This brings us to a major problem: mainstream (conventional) science itself.

We have more than enough proof that UFOs are real, and that they enter and exit our physical world. The problem is that conventional science won't examine the evidence, let alone accept it. Why? There are many reasons for this.

Part of the answer is that conventional scientists have always believed, unquestioningly, that they know nearly all there is to know about Nature and the laws of physics.

Ancient scientists like Claudius Ptolemaeus (Ptolemy), for example, believed that the Earth is the center of the Universe and that the Sun, the planets, and even the stars, revolve around it (a view called "geocentrism").

The Catholic Church agreed, and felt so strongly about this belief that anyone who dared to disagree with it was either tried and imprisoned for life (like Galileo in 1633), or tortured and burned at the stake (like Giordano Bruno in 1600).

But ancient scientists were dead wrong.

Today we know, as Nicolaus Copernicus bravely asserted in the 16th Century, that our Earth is not anywhere close to the center of the Universe; that the Earth and the other eight planets all revolve around the Sun (called "heliocentrism"); and that the millions of stars we see in the night sky are all suns in their own right, with their own solar systems and their own revolving planets.

UFO FACTOID
Five-hundred years ago the Catholic Church arrested, imprisoned, tortured, and even killed people for believing things that today's scientists take for granted. Where will science be 500 years from now?

Since then we've also learned that our solar system itself is located in an out-of-the-way corner in the Milky Way Galaxy, which is only one galaxy of hundreds of billions in the known Universe, a Universe that we now know is infinite in every direction.

Luckily, not every ancient scientist was a "know-it-all," otherwise we wouldn't have progressed very far in the past thousand years. We might still be living in a world without electric lights, flat-screen TVs, cell phones, text messaging, microwaves, computers, cars, jets, and indoor plumbing and heating. Can you imagine reading this book by candlelight? If some early close-minded scientists had had their way, you would be.

Lord Kelvin (William Thomson), a Victorian English scientist, typified false science, which hinders the forward movement of humanity at every step. Here, for instance, is a comment he made in 1899:

> Radio has no future. Heavier-than-air flying machines are impossible. X-rays will prove to be a hoax.

Wouldn't he be surprised if he were alive today?

Despite the obvious lessons learned from Lord Kelvin, many modern day scientists continue to embrace his attitude, particularly in regard to UFOs and their out-of-this world occupants. But why the hubris? Especially when it's clear that there's so much that science still doesn't understand.

Earth's magnetic poles reverse polarity every so often. Why? Conventional scientists have no idea. They also don't know what the Universe is made of, how life originated on Earth, or how we humans store and retrieve memories. Scientists don't even know why cats purr. Yet they believe they know enough to tell us that the unidentified flying objects seen by millions of people the world over are imaginary.

This type of close-minded arrogance is the opposite of real science. Why? Because real science is an open-ended search for truth about everything that touches human existence.

The late Dr. James McDonald, senior physicist at the Institute for Atmospheric Physics, was an example of a true scientist. Here's what the professor had to say about UFOs: "I spent three years personally studying unidentified flying objects and interviewing over 500 American eyewitnesses. My conclusion is that not only are UFOs not illusions, as so many other scientists call them, they are something that every scientist worth his or her salt should take a serious interest in."

Authentic science begins without preconceived notions, asks hard questions, researches every angle, carefully studies the data, reviews the evidence, and accepts the conclusion of its study—no matter what that conclusion turns out to be. This is called the "Scientific Method," and it's used by all good and true scientists.

> UFO FACTOID
> J. Allen Hynek began his scientific career as a skeptic, but later became a believer in UFOs as the overwhelming evidence mounted. Hynek was a true scientist.

False science approaches its search for truth very differently, with both a bias and an agenda, each meant to maintain its current theories and views at all costs. Because of this, false scientists often ask only those questions that will provide answers that support their beliefs, while at the same time they reject opposing views, approach their subject or test with personal assumptions, ignore important evidence, and manipulate the data until the results are what they want them to be.

This is the opposite of both true science and the Scientific Method.

Scientists who take UFOs seriously and study them with an open mind are practicing true science. Scientists who laugh at UFOs and refuse to examine the evidence are practicing false science.

True science accelerates the advancement of knowledge and understanding. False science impedes the advancement of knowledge and understanding.

Here's what one famous authentic scientist, J. Allen Hynek, had to say on the matter in the 1950s: "UFOs cannot be disregarded any longer. We now have empirical evidence of their movements, shapes, and colors. Scientists need to remember that just as there is now a 20th-Century science, one day there will be a 30th-Century science, one that will encompass concepts we can scarcely imagine. We can only reach this stage, however, if we take off our blinders and stop pretending we know everything."

In addressing his close-minded scientific colleagues, Hynek wrote: "I asked my friends in the scientific community to stop poking fun at UFOs and the people who have seen them. And furthermore I told them that UFOs provide us scientists with a golden opportunity to show the public how the scientific method works, particularly in relation to UFOs. In doing so it will be seen that derision is merely disrespect, and neither have any place in true science."

We say well put sir!

SCIENTISTS ONCE LAUGHED AT THE IDEA OF GORILLAS, GIANT SQUID, & METEORS

Let's look at a few intriguing examples of how mainstream science has hampered the progress of science.

> UFO FACTOID
> Scientists once laughed at stories about the gorilla, sneering that it was nothing more than an "imaginary" creature.

Only twenty-five years ago scientists were very comfortable with their view of the Universe. After all, their knowledge was based on the accumulation of thousands of years of scientific thought, theories, and discoveries by some of the most brilliant men and women who've ever lived.

But all of that changed with the discovery of black holes. When one enters the singularity (center) of a black hole one enters a realm of "infinite density," where gravity is so strong that even light can't escape, and where time and space themselves are distorted. Einstein taught us that nothing can travel faster than light, but this idea has been completely shattered by the existence of black holes.

Now scientists have to re-think the way they see the Universe. Why? Because black holes behave in ways that break the laws of Nature as science currently understands them.

UFO FACTOID
Sea monsters are real, and so are meteorites. But at one time scientists arrogantly dismissed both of them as "impossible."

When Europeans first began exploring Africa in the 1800s they brought back bizarre tales of wild "ape-men." Scientists ridiculed the explorers, refusing even to look at physical and photographic evidence (explorers had skins and pictures to prove what they had seen). Scientists believed that no such animal could exist because the very idea violated their long-held cherished beliefs. Today we have a name for the unlikely creatures these daring explorers saw: gorillas.

Scientists once scoffed at the idea of the "sea monster." Reported by sailors and whalers all the way back to ancient times, it was said to be an enormous and hideous animal some seventy feet long, possessing dinner plate-size eyes and a terrible beak-like mouth, with eight arms and two tentacles so long and muscular that it could sink a ship and devour its occupants. Again scientists sneered, stating that "such an animal cannot exist!"

But they stopped laughing when the real thing, a giant squid (*Architeuthis dux*), was "discovered" in the late 1800s. There has been even less laughter since an even bigger squid, the colossal squid (*Mesonychoteuthis hamiltoni*), was discovered in 1925.

UFO FACTOID
Scientists told us that the coelacanth disappeared some 65 million years ago. They ridiculed those who claimed they still existed—until 1938, when a very live coelacanth was hauled in by a South African fisherman.

Scientists also once arrogantly dismissed the idea of **meteors**, thinking they were just ancient myths and silly legends. "How can large stones fall from the heavens?" they joked.

But everyone else knew they were real because they saw their fiery trails blaze through the sky, and because they found the strange objects in their fields while plowing.

American scientists in particular ignored hundreds of eyewitness accounts and refused to inspect the evidence (heavy chunks of unearthly material we now call **meteorites**).

Centuries passed.

In the 1700s European scientists finally "discovered" that meteors do exist, but American scientists continued to disbelieve.

Then in 1807 a large meteor called the "Weston Fireball" exploded over Connecticut, sending down an enormous shower of meteorites. Stunned scientists grudgingly collected over 300 pounds of the strange metal-like fragments and took it back to their laboratories. Forced to open their eyes at long last, they finally accepted what everyone else had known for centuries: meteors and meteorites are real.

UFO FACTOID
Many scientists believe that there can't be anything in the Universe that they don't already know about. And since they don't know about UFOs, then UFOs must not exist. How's that for logic? Luckily for us not all scientists embrace this foolish rule. If they did we'd still be living in a world without electricity, lights, TVs, telephones, microwaves, computers, cars, or jets.

And who could forget the coelacanth? Close-minded scientists once promised us that this primitive fish became extinct 65 million years ago, at the end of the Cretaceous Period. That's a long time ago.

Imagine their surprise when a live coelacanth turned up in a fishing net off the coast of South Africa in 1938. And many more living coelacanths have been captured since then, making a mockery of false science and the false scientists who practice it.

MAINSTREAM SCIENCE VERSUS UFOS

It's obvious from such examples that much of what scientists now believe is scientific "fact" will in the future be considered as outdated as the abacus. And this is a problem of course, because most scientists like neat theories that never change. This brings us back to the subject of UFOs.

Nonbelievers, skeptics, debunkers, and mainstream scientists prefer an orderly Universe that agrees with their own personal worldview of how things should be. Unfortunately UFOs are a square peg that doesn't fit into this round hole. In fact, like black holes, UFOs break almost every known law of physics.

But instead of trying to develop more accurate methods of understanding the UFO phenomenon, mainstream scientists reject it completely, ignore the mountains of evidence available, and treat UFOs as if they're a figment of people's imaginations.

However, as we've seen, this isn't science. It's arrogance. For they're violating one of the fundamental laws of science itself, which is to search for the truth—no matter what the truth may turn out to be—with an open questioning mind. As the famous scientist Gregory Bateson once observed: "It's not science's job to prove, only to probe."

Indeed, scientists are never supposed to say, "we've found the answer, and it's the only answer, and will always be the only answer." Science is supposed to be, in a word, provisional; that is temporary. Why is this?

Science was originally designed to be an open-ended, self-correcting process, one that constantly questions and reevaluates it's own theories, findings, and conclusions. This is why true scientists try to disprove their own theories before announcing them to the public. And this is why nothing in science can ever be considered 100 percent definite. *Science should always be provisional.*

UFO FACTOID
In 1710 the Flemish artist Aert de Gelder created a painting he called "The Baptism of Christ." In it de Gelder clearly depicts a hovering disk-shaped object projecting four beams of light down on John the Baptist as he gently lowers Jesus into the water. Skeptics have no satisfactory explanation for this object. But the rest of us would call it exactly what it is: a UFO.

I will never forget my fourth-grade teacher, who proved, by drawing out long and elaborate equations on the classroom chalkboard, that 2 + 2 actually equals 5. The lesson? Science is fluid and open-ended, not hard and closed.

The popular view of science as nothing more than hard unchanging facts and cold impersonal numbers is incorrect. Real science is actually a blend of both healthy skepticism and creative imagination.

This means that authentic science is 50 percent mysterious art and 50 percent rock-solid data. But false science is always 100 percent of only one or the other.

Skeptics, debunkers, and nonbelieving traditional scientists would be wise to remember the words of one of the world's greatest scientists, Albert Einstein, who, in 1930, wrote:

> The most beautiful thing we can experience is the mysterious. It is the source of all true art and science.

According to Einstein, both art and science are rooted in "the mysterious"; that is, something we can't fully understand. How many scientists today approach science from this viewpoint?

UFOS: ONE OF SCIENCE'S GREATEST CHALLENGES

We've seen the main reason conventional scientists refuse to probe into the riddles of UFOs: UFOs conflict with their comfortable view of the Universe. But there are other reasons as well.

Often individuals who enter the field of science do so because, in part, they have a deep fear of the unknown. Studies show that the types of people who become scientists are usually the kind that prefer the tidy organized explanations that mainstream science provides for the Universe's many mysteries.

> UFO FACTOID
> One study found that scientists would be the most angry of any kind of person at finding out that UFOs and aliens are real.

This type of person likes things to be harmonious, simple, and precise, and would rather not have to deal with messy uncertainties. But UFOs are a messy uncertainty indeed; one that makes complete chaos out of mainstream science's well-ordered world.

It's little wonder then that in a 1961 study entitled "The Implications of a Discovery of Extraterrestrial Life," the Brookings Institute found that of all the different types of people, scientists are the ones who would be the most upset by the discovery of a superior race of beings. Why? Because science is primarily concerned with Man's ability to conquer and control Nature. The existence of UFOs and aliens, however, proves that we are not in complete control of our world. Far from it!

Even the normally scientifically mainstream *Encyclopedia Britannica* admits that:

> The [negative] reaction of physical scientists [toward UFOs] has been centered on reluctance to change their systems of belief based on current physics and astronomy.

While some scientists can't bring themselves to admit that there are creatures in our universe infinitely smarter than we humans, in the end most are afraid of UFOs for one reason and one reason only: they don't know how to explain them. As the Senior Research Officer of France's **CNRS**, Dr. Pierre Guérin, says, "mainstream, nonbelieving scientists are embarrassed by, and even enraged about, UFOs because they simply don't understand what they are."

UFO FACTOID
Conventional science has much in common with conventional religion: both provide simple answers to life's many complexities. This works fine for religion, which is a personal and subjective matter. But it doesn't work well for authentic science, which is supposed to be impersonal and objective.

Instead of admitting this, such scientists downplay, ignore, or make fun of the topic in an attempt to draw attention away from their own ignorance.

This doesn't mean that UFOs aren't real though. It just means that the type of science practiced by nonbelieving scientists is not real science. It's a deeply held personal mythology based on their own theories, views, and fears. This is the opposite of true science.

And where does such an attitude lead? Since these kinds of scientists are not open to the new, have no interest in the unknown, and show no enthusiasm for the unexplained, it can't lead anywhere. This is "dead end science," a type of science that really shouldn't be called science at all.

Skeptics could learn a thing or two from one of Russia's leading scientists, Dr. Felix Zigel, who noted that "scientists who deny the existence of UFOs are behaving irresponsibly. UFOs are a challenging problem, one that needs to be solved, not ignored," Zigel cautioned. "Who better than scientists to unravel this mystery? In fact, it is the God-given duty of every true scientist to try and answer this question and set down the facts once and for all," he staunchly maintained.

General Lionel Max Chassin, commanding general of the French Air Forces in the 1940s, also had something to say about nonbelieving skeptical scientists, engineers, and astronomers. According to General Chassin, "such men and women are so preoccupied with their own omniscience that anything contradicting their personal beliefs can only infuriate them. If they don't know the answer to something they cast doubt on the subject rather than on themselves, disregarding the most obvious evidence in order to do so."

"This overblown pride," Chassin goes on to say, "a type of hubris that was supposed to have died with the likes of Medieval scientists like Galileo and Copernicus, makes false scientists a menace and a hindrance to true science. As so many highly trained and educated people have seen UFOs—and even state that they pose a threat to humanity—the arrogance of nonbelieving scientists has now become not just offensive, but truly dangerous."

> UFO FACTOID
> Arthur Schopenhauer understood that truth is usually rejected before it's accepted.

Thankfully, some astronomers who have seen UFOs have come forward, helping to promote scientific UFO research; brave individuals like Jose A. Y. Bonilla, Edward W. Maunder, Frank Halstead, Walter N. Webb, Dr. Seymour Hess, Dr. Clyde Tombaugh (discover of the planet Pluto), H. Percy Wilkins, Dr. Bart Bok, and Dr. H. Gollnow.

We can be sure that one day UFOs will be accepted by *all* mainstream scientists. Until then let's remember the words of the 19th-Century German philosopher Arthur Schopenhauer:

> All truth passes through three stages: First it is ridiculed; Second, it is violently opposed; and third it is accepted as self-evident.

CHAPTER 10 AT A GLANCE

- Many people reject the idea of UFOs because they go contrary to what conventional science teaches is possible. These individuals are called "skeptics," or "nonbelievers."
- People who want to prove that UFOs don't exist are called "debunkers."
- Skeptics and debunkers feel that people see what they want to see, and that UFOs are nothing more than the products of hoaxes, hallucinations, or hysteria.

- The truth is that most of the people who witness UFOs didn't want to see them, or didn't believe in them to begin with; only 1 percent of all UFO sightings are hoaxed; and mass hallucination and mass hysteria can't explain sightings such as those witnessed over Mexico City, where UFOs are seen by millions of people at the same time.
- Scientists once laughed at the idea of gorillas, meteors, sea monsters, black holes, and the coelacanth—only to later repent of their arrogance.
- Some people refuse to believe that UFOs are real because they seem so unreal.
- Most skeptical scientists refuse to accept the reality of UFOs because they don't know how to explain them.
- Real scientists are very interested in UFOs.
- It's the responsibility of scientists to study UFOs.

11

BECOMING A UFO WATCHER

"The least improbable explanation is that these things, UFOs, are artificial and controlled. My opinion for some time has been that they have an extraterrestrial origin." — Dr. Maurice Bilot, aerodynamicist and mathematical physicist, 1952

We now know what UFOs and aliens are, we've studied their history, and we've examined the strange and often incomprehensible phenomena associated with them. We've also learned what true science is and isn't, and how important it is to have a skeptical but open mind when it comes to UFOs.

In this chapter you'll learn how to become a UFO watcher, what you need to do in order to see a real UFO, and what you should do if you're lucky enough to see one.

BINOCULARS, CAMERAS, & PATIENCE: THE ESSENTIALS OF UFO WATCHING

So you want to see a UFO. What do you do? How does one become a UFO watcher?

To begin with you must look up into the sky. Simply get out of your house, go outside, and look up. This seems obvious. But many people never look into the sky, and yet they complain that they never see UFOs.

So this is the very first thing you must do if you want to see a real unidentified flying object. The more time you spend observing the sky, the better the chance that you'll eventually see a UFO.

Not all UFOs are in the sky of course. Many are seen sitting on the ground, hovering above the ground, or drifting through the canopies of forests. Some are even found underwater in rivers, lakes, and oceans.

UFO FACTOID
A telescope can help you see UFOs in great detail, while a good set of binoculars is essential to becoming a proficient UFO watcher.

The point is to get outside and look around. Because no UFO was ever spotted inside a house, apartment, or building.

When you finally do see a UFO it will be very exciting. It may be the most remarkable experience you'll ever have. But UFOs are often so high up or so far away that they can't be seen clearly.

This is why you should carry a set of binoculars or a telescope with you at all times when you're UFO watching. These instruments will help you zero in on the object and pick out details that you could never see with the naked eye.

Three other pieces of equipment are essential for a UFO watcher's knapsack: a flashlight, a notebook to write down your impressions, and a sketch pad for drawing pictures of what you've seen.

Serious UFO watchers add at least one other item to their surveillance gear: a high resolution still camera and/or video camera (with a tripod if possible). You should always try to record your UFO sighting on film, digital memory card, or tape. Why?

First, you'll want something to prove to others (and even to yourself) that what you've seen is not a hallucination, that it's real.

Second, the police, or even the military, may want to look at what you have.

Third, local news crews will be eager to air your images on TV.

And fourth, magazines, newspapers, and film companies may even be willing to pay you to use what you've captured. The *National Enquirer*, for example, offers a prize of 1 million dollars to anyone who can show positive proof of a real UFO.

UFO FACTOID
Always carry a notebook with you while UFO hunting, so you can record your impressions.

Another important thing for a UFO watcher to have is patience. Most people have to spend many hours, days, or even months, gazing up into the sky before seeing a UFO. Still others spend their entire lives looking and never see one.

So you must be patient if you're ever to catch a glimpse of one of our Universe's most astounding and beautiful mysteries.

One last thing: good UFO watchers are always responsible, polite, courteous, and sensitive to others. Don't trespass on private property without the owner's permission, and always pick up after yourself. Always leave your UFO watching site even cleaner than it was when you arrived.

> UFO FACTOID
> A professional UFO watcher never leaves home without a camera, flashlight, and cell phone.

And if you're in a UFO watching group, don't shine your lights in other people's faces. Help beginners with their equipment. Freely share your knowledge about UFOs and aliens with other members. And always respect the rights, the privacy, and the personal belongings of others.

UFO HOT SPOTS: HIGH TECH FACILITIES

Although, as we've learned, UFOs can be seen anywhere at anytime, there are certain places where UFOs seem to gather, and where there's a better chance of spotting one. These places are called UFO "**hot spots**."

Some UFO hot spots seem to be unrelated to anything specific, such as the regions around Norwich, England, and Bonnybridge, Scotland, where, mysteriously, more and more UFO sightings seem to be occurring each year (in Bonnybridge alone literally thousands of UFO reports have been filed since 1990).

But most UFO hot spots are directly related to specific places or features. These can be broken down into two categories: the high-tech facility and the sacred site.

No doubt the hottest UFO hot spot of the high-tech facility variety is the military base. Studies of UFO reports show that more UFOs are seen in the area of military bases than anywhere else on Earth. Why?

Aliens seem fascinated by our war technologies: planes, jets, bombs, tanks, drones, and nuclear missiles. Maybe this is because they're monitoring the development of our weapons systems and combat equipment for a possible future invasion against us. Or, more likely, they're concerned that we may eventually destroy ourselves, and they want to try and prevent this from happening.

> UFO FACTOID
> Military bases are considered the number one UFO hot spot.

So just by looking for UFOs near a military installation you'll increase your odds of seeing one by many times.

Casebook Study 61: One example of UFOs' interest in our weapons came on September 15, 1964, at Vandenberg Air Force Base, California. On this day scientists sent up an Atlas missile on a trial run. A telescope on the ground was fixed with a movie camera that was recording the entire event.

Unfortunately, when the missile reached a height of sixty miles above the Earth's surface it began to spin out of control, crashing into the Pacific Ocean.

Later when the film was played back scientists were shocked to see a circular UFO fly right up to their missile as it streaked through the air. The object flew easily around it, setting off bright flashes of light as it went. The UFO then hovered momentarily above the missile, and disappeared. Seconds later the missile malfunctioned and plunged into the ocean.

If UFOs have the ability to control our weaponry, is there really any limitation to their powers?

Airports, seaports, electric and nuclear power plants, and nuclear development facilities, are also magnets for UFOs. Again, it seems that aliens are extremely attracted to human-made machines and to our energy-producing technologies.

UFO FACTOID
Aliens are fascinated with our spacecraft and weapons systems, and with good reason.

UFOs have often been seen pacing airplanes and passenger jets that are taking off or landing at airports, and we have many reports of them "snooping" around power plants of all kinds. The Los Alamos Atomic Energy Commission Project, at Los Alamos, New Mexico (the site of the development of the world's first atomic bombs), is one of the more popular UFO hot spots in this category.

Another high-tech hotspot is the space flight launching facility, such as **NASA**'s Kennedy Space Center (**KSC**) at Cape Canaveral, Florida. NASA has had many run-ins with UFOs over the years, such as when one pursued a Polaris missile over the Atlantic Ocean in January 1961, and when one chased a Gemini space capsule in April 1964.

There are space flight launching facilities all over the world. If you're fortunate enough to live near one you may also be fortunate enough to see a UFO.

OTHER UFO HOT SPOTS: SACRED SITES & HOLY GROUND

A place can also become a UFO hot spot because of something that has occurred there in the past. Perhaps a UFO crashed in that spot, or maybe some kind of bizarre military activity took place there.

UFO FACTOID
Vortex centers, portals between the supernatural and the natural planes, are also UFO hot spots, perhaps for this very reason.

Whatever the reason this type of UFO hot spot becomes "hot," UFO enthusiasts and ufologists think of them in exactly the same way religious people view a temple, synagogue, or church: as "holy ground."

Let's take a look at some of the more popular sacred UFO hot spots.

- *Roswell, New Mexico*: Arguably the most famous piece of UFO holy ground in the U.S., the Roswell Incident was the first time in history that a government openly and publicly admitted that UFOs are real. It's also the site of the crash of a genuine flying saucer—with alien occupants.

- *Rendlesham Forest, Suffolk, England*: Though the military base (RAF-USAF Bentwaters-Woodbridge) that once stood here has now been shut down, the forest where the 1980 UFO landed still stands as a sacred monument to one of the greatest alien encounters in human history.

- *Sand Dunes State Forest, Minnesota*: After a major UFO flap here in 1992, local ufologists and believers began gathering each night in an attempt to intentionally initiate contact with aliens.

Sedona, Arizona: This area is known to be a "**vortex center**": a point of contact between the physical world and the spiritual world. Native Americans have recognized the mystical vibrations of the Sedona region for thousands of years and believe that the Earth here gives off special healing and magical energies.

Not surprisingly, many UFOs have been sighted at Sedona, making it one of the more intriguing sacred UFO hot spots. (Other vortex centers include: Stonehenge; the Great Pyramid; Machu Picchu; Newgrange, Ireland; Easter Island; Angkor Wat; the North Pole; the South Pole; and Malta.)

Gulf Breeze, Florida: From the 1980s to the present day this area has proven to be one of the best places in the world to see UFOs. In part this is due to the long low-lying beaches that give UFO watchers an excellent view of the sky. Another reason is the military base situated in the region.

Dulce, New Mexico: Ufologists believe that somewhere beneath the Archuleta Mesa at Dulce lies a huge, 4.8 mile wide, secret underground facility where medical experiments are being performed cooperatively by both humans and aliens.

Known as "Dulce Base," many have come forward with stories of hideous reptilian ETs called Dracos, who work in the lowest of seven levels, known as "Nightmare Hall," where horrific genetic tests are taking place.

Besides eyewitness testimony from both former base employees and local residences as to the facility's existence, mystery helicopters, cattle mutilations, and even UFOs have been seen around the mesa dating back several decades.

Boulia, Australia: For hundreds of years Australians from around Boulia have been seeing what's called "the Min Min Light." This strange object appears at night as a small bright blue and red oval light that hovers, flickers, and dances in the night air, inviting people to follow it.

Many have tried doing just that, but so far no one has ever even gotten close to catching it. In fact, a number of Min Min Light hunters have gotten in car accidents, or even died, pursuing the eerie object.

Despite the dangers, tourists flock from far and wide to this UFO hot spot to catch a glimpse of the Min Min Light for themselves.

UFO FACTOID
To keep out prying eyes, the U.S. military is buying up more and more land around Area 51. At present the facility can only be seen through an extremely high-powered telescope from a mountaintop some twenty miles distant. Eventually even this area will be sealed off to the public.

Giant Rock, California: Ever since a man encountered an alien here in 1953 this ancient Native American sacred site has become a UFO hot spot.

Groom Lake, Nevada: The site of Area 51—the American military base that the government claims "doesn't exist"—is a region that's becoming hotter and hotter in recent years, no doubt because of the base's reputation for secretly studying and flying genuine extraterrestrial craft. Another reason is the many weird dancing lights that witnesses have filmed flying over the mountains here.

Yakima, Washington: Another sacred Native American site, Yakima, on the border between Oregon and California, is a vortex center where UFOs seem to enter the physical realm from another dimension. From the 1960s onward hundreds of strange UFOs have been seen here, from brilliant green, orange, and red pulsating lights, to huge fireballs that chase cars. Aliens have also been encountered in the Yakima region, as has Bigfoot. And many think there's a connection.

Mount Rainier, Washington: This area, where Kenneth Arnold had his famous June 1947 sighting of nine shiny flying disks, is where the entire modern UFO era began. Because of this Mount Rainier is considered one of the holiest of all UFO sites, and believers still gather here every June to celebrate the occasion.

WHAT TO DO DURING AND AFTER YOU SEE A UFO

If you were visiting one of the world's many UFO hot spots, or even if you were in your own backyard, and you spotted a UFO, what would you do?

Unfortunately most people who see UFOs are so overwhelmed by the experience that they don't do anything at all. They don't even report them. This is poor UFO watching.

A good UFO watcher takes the sighting of a UFO very seriously, and they treat ufology itself professionally, objectively, respectfully, and scientifically.

Even while the sighting is taking place he or she will always try to keep calm and use common sense. This is because they know that they'll need to remember as much information as possible so that they can record (and report) the sighting accurately later on.

For your UFO report to be valuable to science there are a number of things you should do during and immediately following your sighting.

> UFO FACTOID
> Approach all UFO sightings seriously, as a true scientist. Other people will want to study the data you collect in order to learn more about the UFO phenomenon.

During the sighting:

- Look around for anyone else who might also be seeing the UFO, or for someone who could be a possible witness. The more eyewitnesses the better.

- Next, if you have a camera take a picture of the object. As many as you can. If you have a video camera, record it. Record as long as you can. Whether you photograph or video the UFO, try to get trees, buildings, or telephone poles (that is, anything manmade) in the frame, for reference.

- Whether you do or you don't have a way to photograph or film the object, get out your sketchpad and draw a picture of it. Be careful to include as many details as possible.

If the sighting lasts long enough, call the police. Then call a local TV station to see if they can get a news crew to the site in time to film it themselves.

After the sighting:

Once your sighting is over take out your notepad or notebook (real or electronic) and record everything you can remember. For example, the color(s), shape, size, and behavior of the craft, time of day or night, air temperature, the direction it moved, and its speed.

> UFO FACTOID
> Let the police know about your sighting. If there's physical evidence, wait for them to arrive at the scene before you touch anything.

If the craft actually touched the ground, or left behind anything such as angel hair, broken tree limbs, impressions, or burned and flattened grass, don't touch anything. Photograph or video the area, then protect it until the police arrive.

In your notebook describe how you felt as you watched the UFO.

Don't forget to record the names, addresses, and telephone numbers of the other witnesses, if there were any.

Call or contact a UFO organization by phone or via the Internet as soon as you can. You can submit an online UFO sighting report, for example, with MUFON ("Mutual UFO Network") in Newport Beach, California. Their Website address is: www.mufon.com.

Another place to report your sighting is CUFOS (the "Center for UFO Studies"), in Chicago, Illinois. Their Website address is: www.cufos.org.

Both organizations treat UFOs in a scientific manner, so you can be sure your sighting will be handled with the seriousness it deserves.

If you prefer calling a telephone hotline, there's The National UFO Reporting Center in Seattle, Washington. Their telephone number is: 206-722-3000. Website: www.nuforc.org.

Contact the news department of a local radio station, TV station, and/or newspaper. A good UFO watcher has all of these telephone numbers written down in their notebook or stored on their cell phone beforehand so that they can get at them quickly.

> UFO FACTOID
> Are we alone? The Universe contains hundreds of billions of galaxies, each one itself containing billions of solar systems like ours. This means that planets probably number in the trillions. Many, like our Earth, are capable of sustaining life. As a result of this knowledge even the most hard-headed skeptical scientists today believe that there are other lifeforms somewhere in the Universe.

THE UFO SIGHTING REPORT FORM

There's only a one in ten chance that your sighting will turn out to be a genuine UFO. More likely it will be a natural object (such as the planet Venus or a meteor), or a human-made object (such as a plane or a blimp).

You can help investigators determine whether your sighting was a real UFO or not by filling out the UFO sighting report form located in Appendix A. (Following it, Appendix B, is a sample of how your report might look after you complete it.)

Make photocopies of this blank form and carry it with you while you're UFO watching. If you see something unexplained, fill one of them out and give it, or mail it, to the proper authorities.

CHAPTER 11 AT A GLANCE

- The more time you spend observing the heavens the more likely it is that you'll see a UFO. If you're near or on a river, lake, or sea, be sure to scan the depths and surface of the water as well.
- Becoming a good UFO watcher means carrying essential equipment with you.
- Carry binoculars or a telescope in order to see the details of the object.
- You'll need a real or digital notepad to record a description of the object.
- You'll need a sketchpad to draw a picture of the object.

- Take a cell phone camera, regular camera, or a video-recorder with you when you go UFO watching. You'll want some kind of photographic evidence of what you've seen.
- Good UFO watchers are patient. UFOs appear when they want to, not when we want them to.
- A good UFO watcher is courteous, respectful, and responsible.
- Though UFOs can be seen anywhere anytime, they sometimes prefer certain places over others, such as Bonnybridge, Scotland. These places are called UFO "hot spots."
- One type of UFO hot spot is the high-tech facility. This would include: military bases, airports, seaports, electric and nuclear power plants, and space flight launching facilities.
- Another type of UFO hot spot is the sacred site. This would include such places as: Roswell, New Mexico; Mount Rainier, Washington; and Gulf Breeze, Florida.
- Good UFO watchers take their hobby very seriously, and every sighting is treated professionally and scientifically.
- During a UFO sighting you must see if there are other witnesses, take photos or video of the object, and (if the object is visible for a long time) call the police, then a local TV station.
- After the sighting write down everything you can remember in your notebook.
- Fill out a copy of your UFO Sighting Report Form (Appendix A).
- If the object landed on the ground seal off the area until the police arrive. Photograph the area.
- Draw a picture of the object on your sketchpad.
- Call a UFO sighting report center, or contact UFO groups and Internet UFO sites.
- Contact local newspapers and radio stations.

BECOMING A UFOLOGIST

"Army intelligence has recently said that the matter of 'Unidentified Aircraft' or 'Unidentified Aerial Phenomena,' otherwise known as 'Flying Discs,' 'Flying Saucers,' and 'Balls of Fire,' is considered top secret by intelligence officers of both the Army and Air Forces."
— FBI memo, 1949

SO YOU WANT TO BE A UFOLOGIST?

Over the last eleven chapters we've learned a great deal about UFOs and aliens. In fact, we've touched on all the basics of this most fascinating subject—which is why this book is subtitled, "The Complete Guidebook."

But even so, the information here is just the tip of the iceberg. There is much more to learn, far more than could ever be put in one book, no matter how big it is. Every writer, scientist, ufologist, and UFO enthusiast in the world has something unique to contribute to the ever growing body of UFO knowledge.

> UFO FACTOID
> The surface of Mars contains pyramids, sculptures, tunnels, building compounds, and triangles, as well as numerous other types of complex artificial features and monuments. How did they get there? As of 2015 humans have yet to step foot on the Red Planet.

If you've enjoyed what you've learned so far and want to know more, you might want to consider becoming a ufologist. Just by reading this book you've already become a budding UFO expert.

But how do you know if it's right for you?

Everyone who becomes a ufologist has one thing in common: any mention of the words "UFO" or "aliens" brings about an instant sense of elation and happiness we call **UFOria** (a play on the Greek-Latin word *euphoria*).

Of course, one doesn't go to school to become a ufologist, and there's no such thing as a "degree in ufology." All ufologists are self-taught; that is, they study and learn on their own, and with the help of others.

So if your goal is to become a ufologist, read on. This chapter will give you the tools you need to get started. Stick with it and follow the guidelines. By the time you're through you'll be well on your way to becoming a real ufologist.

> UFO FACTOID
> While there are not yet any schools devoted to ufological studies, MUFON offers workshops for training individuals to become UFO field investigators.

BOOKS ABOUT UFOS AND ALIENS

The first step to becoming a ufologist is to read and study every book you can on the topic of UFOs and aliens.

You can take them out at your local library, buy them at a bookstore or online, or borrow them from friends or other people involved in UFO research. As you go along, keep track of the books you've read, the authors, the titles, and especially what you've learned from each one. Don't forget to recommend them to your friends, family members, and fellow UFO watchers.

Below is a list of books to help get you started, listed by category and genre. All are directly or indirectly related to UFOs and aliens. (To provide a well rounded view I've also included books by skeptics.)

UFOS (GENERAL)

- *UFOs and Aliens: The Complete Guidebook*, Lochlainn Seabrook.
- *Open Skies, Open Minds: For the First Time a Government UFO Expert Speaks Out*, Nick Pope.
- *The Phoenix Lights: A Skeptic's Discovery That We Are Not Alone*, Lynne D. Kitei.
- *The UFO Experience: A Scientific Inquiry*, J. Allen Hynek.
- *The Report on Unidentified Flying Objects*, Edward J. Ruppelt.
- *The Y-Files: UFO Encounters From Across Yorkshire*, Simon Ritchie.
- *UFOs: Interplanetary Visitors*, Raymond E. Fowler.
- *Casebook of a UFO Investigator*, Raymond E. Fowler.
- *The UFO Controversy in America*, David M. Jacobs.

- *Alien Update*, Timothy Good (editor).
- *One in Forty: The UFO Epidemic*, Preston Dennett.
- *The Mystery of the Ghost Rockets*, Loren Gross.
- *UFOs and Related Subjects: An Annotated Bibliography*, Lynn Catoe.
- *The UFO Encyclopedia*, Margaret Sachs.
- *Incident at Exeter*, John Fuller.
- *UFOs: Operation Trojan Horse*, John Keel.
- *Flying Saucers on the Attack*, Harold T. Wilkins.
- *UFO Exist!*, Paris Flammonde.
- *The Age of Flying Saucers*, Paris Flammonde.
- *UFOs From Behind the Iron Curtain*, Ion Hobana and Julien Weverbergh.
- *UFOs Past, Present, and Future*, Robert Emenegger.
- *The UFO Experience*, J. Allen Hynek.
- *The Reference for Outstanding UFO Sighting Reports*, Thomas Olsen.
- *Mysteries of the Unknown: The UFO Phenomenon*, Time-Life Books.
- *The Gulf Breeze Sightings*, Ed Walters and Frances Walters.
- *Report on UFO Wave of 1947*, Ted Bloecher.
- *Inside the Space Ships*, George Adamski.
- *Flying Saucers Have Landed*, Desmond Leslie and George Adamski.
- *Flying Saucers*, Donald Menzel.
- *The UFO Enigma*, Donald Menzel and Ernest Taves.
- *The Encyclopedia of UFOs*, Ronald Story.
- *UFOs and the Limits of Science*, Ronald Story and J. Richard Greenwell.
- *The World of Flying Saucers*, Donald Menzel and Lyle G. Boyd.
- *UFO Encounters and Beyond*, Jerome Clark.
- *UFOs in the 1980s*, Jerome Clark.
- *The UFO Encyclopedia*, Jerome Clark.
- *The Unidentified: Notes Toward Solving the UFO Mystery*, Jerome Clark and Loren Coleman.
- *The UFO Casebook*, Peter Brookesmith (editor).
- *Flying Saucers: Serious Business*, Frank Edwards.
- *Flying Saucers: Here and Now*, Frank Edwards.
- *The UFO Handbook: A Guide to Investigating, Evaluating and Reporting UFO Sightings*, Allan Hendry.

- *The Case for the UFO*, Morris K. Jessup.
- *The UFO Evidence*, Richard Hall.
- *Flying Saucers and the U.S. Air Force*, Lawrence Tacker.
- *The Coming of the Saucers*, Kenneth Arnold and Ray Palmer.
- *The UFO Encyclopedia*, John Spencer.
- *The Flying Saucer Review's World Roundup of UFO Sightings*, Brinsley Le Poer Trench.
- *The Flying Saucer Story*, Brinsley Le Poer Trench.
- *UFOs & How to See Them*, Jenny Randles.
- *UFO Reality*, Jenny Randles.
- *Science and the UFOs*, Jenny Randles and Peter Warrington.
- *UFOs and Outer Space Mysteries: A Sympathetic Skeptic's Report*, James E. Oberg.
- *The Edge of Reality*, J. Allen Hynek and Jacques Vallee.
- *Over the Southern Hemisphere*, Michael Hervey.
- *The UFO Invasion: The Roswell Incident, Alien Abductions, and Government Coverups*, Kendrick Frazier, Barry Karr, and Joe Nickell (editors).
- *Flying Saucers: The Startling Evidence of the Invasion From Outer Space*, Coral Lorenzen.
- *UFOs Over America*, Jim and Coral Lorenzen.
- *Flying Saucers: Top Secret*, Donald E. Keyhoe.
- *The Flying Saucers Are Real*, Donald E. Keyhoe.
- *Flying Saucers from Outer Space*, Donald E. Keyhoe.
- *Flying Saucers: A New Look*, Donald E. Keyhoe and Gordon I. R. Lore (editors).

EARLY UFO SIGHTINGS (before the 21st Century)

- *The Spaceships of Ezekiel*, Josef F. Blumrich.
- *The Great Airship Mystery: A UFO of the 1890s*, Daniel Cohen.
- *The Bible and Flying Saucers*, Barry H. Downing.
- *Mysterious Beings*, John Keel.
- *The Great Texas Airship Mystery*, Wallace O. Chariton.
- *Dragons*, Peter Hogarth and Val Clery.
- *The Eye*, Francis Huxley.
- *The UFO Wave of 1896*, Loren E. Gross.

- The Airship File: *A Collection of Texts Concerning Phantom Airships and Other UFOs*, Thomas E. Bullard.
- *UFOs: A Pictorial History From Antiquity to the Present*, David C. Knight.

ROSWELL & OTHER UFO CRASHES

- *Flying Saucers and Science: A Scientist Investigates the Mysteries of UFOs: Interstellar Travel, Crashes, and Government Cover-Ups*, Stanton T. Friedman.
- *The Roswell Incident*, Charles Berlitz and William L. Moore.
- *The Day After Roswell: A Former Pentagon Official Reveals the U.S. Government's Shocking UFO Cover-up*, Philip J. Corso.
- *Roswell UFO Crash Update: Exposing the Military Cover-up of the Century*, Kevin Randle.
- *UFO Retrievals: The Recovery of Alien Spacecraft*, Jenny Randles.
- *Crash at Corona: The Definitive Study of the Roswell Incident,* Stanton T. Friedman and Don Berliner.
- *The GAO Report on Roswell*, U.S. General Accounting Office.
- *UFO Crash at Roswell*, Kevin D. Randle and Donald R. Schmitt.
- *Sky Crash*: Brenda Butler, Jenny Randles, and Dot Street.
- *Retrievals of the Third Kind*, Leonard H. Stringfield.
- *UFO Crash / Retrievals: Amassing the Evidence*, Leonard H. Stringfield.
- *The Truth about the UFO Crash at Roswell*, Kevin D. Randle and Donald Schmitt.

ALIENS & ALIEN ABDUCTION

- *The Uninvited: An Expose of the Alien Abduction Phenomenon*, Nick Pope.
- *Captured! The Betty and Barney Hill UFO Experience: The True Story of the World's First Documented Alien Abduction*, Stanton T. Friedman and Kathleen Marden.
- *Abduction In My Life: A Novel of Alien Encounters*, Dr. Bruce Maccabee.
- *Alien Contact: The First Fifty Years*, Jenny Randles.
- *Alien Agenda: Investigating the Extraterrestrial Presence Among Us*, Jim Marrs.

- *UFO Visitation: Preparing for the 21st Century*, Alan Watts.
- *We Are Not Alone*, Walter Sullivan.
- *Encounters with UFO Occupants*, Coral and Jim Lorenzen.
- *Abducted!: Confrontations with Beings from Outer Space*, Coral and Jim Lorenzen.
- *Fire in the Sky: The Walton Experience*, Travis Walton.
- *The Alien World*, Peter Brookesmith (editor).
- *UFO Abductions: The Measure of a Mystery*, T. E. Bullard.
- *The Humanoids: A Survey of Worldwide Reports of Landings of Unconventional Aerial Objects and Their Alleged Occupants*, Charles Bowen (editor).
- *Angels: An Endangered Species*, Malcolm Godwin.
- *The Andreasson Affair*, Raymond E. Fowler.
- *The Andreasson Affair: Phase Two*, Raymond E. Fowler.
- *The Watchers: The Secret Design Behind UFO Abduction*, Raymond E. Fowler.
- *The Allagash Abductions: Undeniable Evidence of Alien Intervention*, Raymond E. Fowler.
- *Missing Time: A Documented Study of UFO Abductions*, Bud Hopkins.
- *UFO Abductions: True Cases of Alien Kidnappings*, D. Scott Rogo.
- *The Tujunga Canyon Contacts*, Ann Druffel and D. Scott Rogo.
- *Visitors from Outer Space*, Roy Stemman.
- *Alien Creatures*, Jean-Claude Suarès and Richard Siegal.
- *Aliens in the Sky*, John Fuller.
- *The Interrupted Journey*, John Fuller.
- *Passport to Magonia: From Folklore to Flying Saucers*, Jacques Vallee.
- *Revelations: Alien Contact and Human Deception*, Jacques Vallee.
- *Dimensions: A Casebook of Alien Contact*, Jacques Vallee.
- *The Invisible College: What a Group of Scientists Has Discovered About UFO Influences on the Human Race*, Jacques Vallee.
- *The Alien Tide*, Tom Dongo.
- *Secret Life: Firsthand Accounts of UFO Abductions*, David M. Jacobs.
- *The Other*, Brad Steiger.
- *The Evidence for Alien Abductions*, John Rimmer.
- *Uninvited Visitors: A Biologist Looks at UFOs*, Ivan T. Sanderson.
- *Angels and Aliens*, Keith Thompson.

- *Light Years: An Investigation into the Extraterrestrial Experiences of Eduard Meier*, Gary Kinder.
- *Abduction: Human Encounters With Aliens*, John E. Mack.
- *Communion: A True Story*, Whitley Strieber.
- *Transformation: The Breakthrough*, Whitley Strieber.
- *The Secret School*, Whitley Strieber.
- *Parallel Universes*, Fred Alan Wolf.
- *Visions, Apparitions, Alien Visitors: A Comparative Study of the Entity Enigma*, Hilary Evans.
- *Intruders: The Incredible Visitations at Copley Wood*, Bud Hopkins.

THE UFO COVERUP

- *Open Skies, Open Minds: For the First Time a Government UFO Expert Speaks Out*, Nick Pope.
- *Top Secret/Majic: Operation Majestic-12 and the United States Government's UFO Cover-up*, Stanton T. Friedman.
- *Above Top Secret: The Worldwide UFO Cover-Up*, Timothy Good.
- *Beyond Top Secret: The Worldwide UFO Security Threat*, Timothy Good.
- *Alien Liaison: The Ultimate Secret*, Timothy Good.
- *Need to Know: UFOs, the Military, and Intelligence*, Timothy Good.
- *UFO-FBI Connection: The Secret History of the Government's Cover-up*, Dr. Bruce Maccabee.
- *The Secret Government: The Origin, Identity, and Purpose of MJ-12*, Milton William Cooper.
- *A Covert Agenda: The British Government's UFO Top Secrets Exposed*, Nicholas Redfern.
- *The UFO Files: The Canadian Connection Exposed*, Palmiro Campagna.
- *The Oz Files: Government Files Reveal the Inside Story of Australian UFO Sightings*, Bill Chalker.
- *Left at East Gate: A First-Hand Account of the Bentwaters-Woodbridge UFO Incident, Its Cover-up, and Investigation*, Larry Warren and Peter Robbins.
- *Project Blue Book*, Brad Steiger (editor).
- *The UFO Cover-up: What the Government Won't Say*, Lawrence Fawcett and Barry J. Greenwood.

- *Underground Bases and Tunnels: What is the Government Trying to Hide?*, Richard Sauder.
- *MJ-12 and the Riddle of Hangar 18*, Timothy Green Beckley.
- *UFOs? Yes! Where the Condon Committee Went Wrong*, David R. Saunders and Roger R. Harkins.
- *Silent Invasion: The Shocking Discoveries of a UFO Researcher*, Ellen Crystal.
- *The Flying Saucer Conspiracy*, Donald E. Keyhoe.
- *Situation Red: the UFO Siege!*, Leonard H. Stringfield.

MEN IN BLACK

- *Casebook on the Men in Black*, Jim Keith.
- *The Truth Behind Men in Black: Government Agents—or Visitors from Beyond*, Jenny Randles.
- *They Knew Too Much About Flying Saucers*, Gray Barker.
- *Flying Saucers and the Three Men*, Albert K. Bender.

ANIMAL MUTILATIONS & MYSTERY HELICOPTERS

- *Mute Evidence*, Daniel Kagan and Ian Summers.
- *Cattle Mutilation: The Unthinkable Truth*, Frederick W. Smith.
- *The Choppers and the Choppers: Mystery Helicopters and Animal Mutilations*, Tom Adams.
- *Mystery Helicopters and Animal Mutilations: Exploring a Connection*, Tom Adams and Gary Massey.
- *Messengers of Deception: UFO Contacts and Cults*, Jacques Vallee.
- *Mystery Stalks the Prairie*, Roberta Donovan.

CROP CIRCLES & VORTEX CENTERS

- *Circular Evidence*, Pat Delgado and Colin Andrews.
- *Crop Circles: The Latest Evidence*, Pat Delgado and Colin Andrews.
- *Earth Magic*, Francis Hitching.
- *The Circles Effect and Its Mysteries*, George Terence Meaden.
- *Circles from the Sky*, George Terence Meaden.
- *The Goddess of the Stones*, George Terence Meaden.
- *The Crop Circle Enigma*, Ralph Noyes (editor).
- *Crop Circles: A Mystery Solved*, Jenny Randles and Paul Fuller.

- *The Mysteries of Sedona*, Tom Dongo.
- *Sedona UFO Connection*, Richard Dannelley.

MYSTERIOUS LIGHTS

- *The Mystery of the Green Fireballs*, William L. Moore (editor).
- *Earth Lights Revelation: UFOs and Mystery Lightform Phenomena: The Earth's Secret Energy*, Paul Devereux.
- *The Marfa Lights: A Viewer's Guide*, Dennis Stacy.
- *Mysterious Fires and Lights*, Vincent H. Gaddis.

BERMUDA TRIANGLE, SEA MYSTERIES, & STRANGE DISAPPEARANCES

- *Into Thin Air: People Who Disappear*, Paul Begg.
- *The Inexplicable Sky*, Arthur Constance.
- *The Disappearance of Flight 19*, Larry Kusche.
- *The Bermuda Triangle*, Charles Berlitz and J. Manson Valentine.
- *Without a Trace*, Charles Berlitz and J. Manson Valentine.
- *Invisible Horizons: True Mysteries of the Sea*, Vincent Gaddis.
- *A Disquisition Upon Certain Matters Maritime, and the Possibility of Life Under the Waters of This Earth*, Ivan T. Sanderson.

UNEXPLAINED MYSTERIES (GENERAL)

- *Carnton Plantation Ghost Stories: True Tales of the Unexplained From Tennessee's Most Haunted Civil War House!*, Lochlainn Seabrook.
- *Unexplained! 347 Strange Sightings, Incredible Occurrences, and Puzzling Phenomena*, Jerome Clark.
- *Parallel Worlds*, Dr. Michio Kaku.
- *Fingerprints of the Gods*, Graham Hancock.
- *The Monuments of Mars*, Richard C. Hoagland.
- *We, From Mars: Old and New Hypotheses about the Red Planet*, Walter Hain.
- *Extraterrestrial Archaeology*, David Childress.
- *The Lost World of Agharti: The Mystery of Vril Power*, Alec Maclellan.
- *The Bigfoot Casebook*, Janet Bord and Colin Bord.
- *The Evidence for Bigfoot and Other Man-Beasts*, Janet Bord and Colin Bord.

- *Creatures of the Outer Edge*, Jerome Clark and Loren Coleman.
- *Bigfoot*, Ann B. Slate and Alan Berry.
- *Abominable Snowman: Legend Come to Life*, Ivan T. Sanderson.
- *Bigfoot, Yeti and Sasquatch in Myth and Reality*, John Napier.
- *The Search for Bigfoot: Monster, Myth or Man?*, Peter Byrne.
- *Exotic Zoology*, Willy Ley.
- *The Sirius Mystery*, Robert K. G. Temple.
- *The Orion Mystery: Unlocking the Secrets of the Pyramids*, Robert Bauval and Adrian Gilbert.
- *Gods of the New Millennium*, Alan Alford.
- *Fire From Heaven: A Study of Spontaneous Combustion in Human Beings*, Michael Harrison.
- *In the Wake of the Sea-Serpents*, Bernard Heuvelmans.
- *The Dragon and the Disc: An Investigation Into the Totally Fantastic*, F. W. Holiday.
- *The Monsters of Loch Ness*, Roy P. Mackal.
- *Loch Ness Monster*, Tim Dinsdale.
- *In Search of Lake Monsters*, Peter Costello.
- *The Loch Ness Monster and Others*, Rupert T. Gould.
- *The Case for Sea Serpents*, Rupert T. Gould.
- *Our Ancestors Came From Outer Space: A NASA Expert Confirms Mankind's Extraterrestrial Origins*, Maurice Chatelain.
- *Somebody Else Is on the Moon*, George H. Leonard.
- *We Discovered Alien Bases on the Moon*, Fred Steckling.
- *The Poltergeist Experience*, D. Scott Rogo.
- *The Haunted Universe: A Psychic Look at Miracles, UFOs and Mysteries of Nature*, D. Scott Rogo.
- *The Legend and Bizarre Crimes of Spring Heeled Jack*, Peter Haining.
- *Curious Encounters: Phantom Trains, Spooky Spots, and Other Mysterious Wonders*, Loren Coleman.
- *Mysterious Canada: Strange Sights, Extraordinary Events, and Peculiar Places*, John Robert Colombo.
- *Strange Mysteries of Time and Space*, Harold T. Wilkins.
- *Mysteries of the Unexplained*, Carroll C. Calkins.
- *Mind Over Space*, Nandor Fodor.
- *Strange Creatures From Time and Space*, John Keel.

- *Thunderbirds!: America's Living Legends of Giant Birds*, Mark A. Hall.
- *The Fire Came By: The Riddle of the Great Siberian Explosion*, John Baxter and Thomas Atkins.
- *The Mothman Prophecies*, John Keel.
- *Alien Animals*, Janet Bord and Colin Bord.
- *Strangers From the Skies*, Brad Steiger.
- *Mysterious Disappearances of Men and Women in the U.S.A., Britain and Europe*, Harold T. Wilkins.
- *An Encyclopedia of Fairies: Hobgoblins, Brownies, Bogies, and Other Supernatural Creatures*, Katharine Briggs.
- *A Living Dinosaur?: In Search of Mokele-Mbembe*, Roy P. Mackal.
- *Chariots of the Gods?: Unsolved Mysteries of the Past*, Erich Von Däniken.
- *The Return of the Gods: Evidence of Extraterrestrial Visitation*, Erich Von Däniken.
- *Mysterious America*, Loren Coleman.
- *Stranger than Science*, Frank Edwards.
- *Strangest of All*, Frank Edwards.
- *Searching for Hidden Animals: An Inquiry Into Zoological Mysteries*, Roy P. Mackal.
- *Oddities: A Book of Unexplained Facts*, Rupert T. Gould.
- *On the Track of Unknown Animals*, Bernard Heuvelmans.
- *Kingdoms Within Earth*, Norma Cox.
- *Subterranean Worlds: 100,000 Years of Dragons, Dwarfs, the Dead, Lost Races and UFOs from Inside the Earth*, Walter Kafton-Minkel.
- *Secret of the Ages: UFOs From Inside the Earth*, Brinsley le Poer Trench.
- *Handbook of Unusual Natural Phenomena*, William R. Corliss.
- *The Mysterious World: An Atlas of the Unexplained*, Francis Hitching.

SPIRITUALITY, FEMININE RELIGION (THEALOGY), RELIGIOUS MYSTERIES, & MYTHOLOGY

- *Britannia Rules: Goddess-Worship in Ancient Anglo-Celtic Society*, Lochlainn Seabrook.

- *The Book of Kelle: An Introduction to Goddess-Worship and the Great Celtic Mother-Goddess Kelle, Original Blessed Lady of Ireland*, Lochlainn Seabrook.
- *The Goddess Dictionary of Words and Phrases*, Lochlainn Seabrook.
- *Aphrodite's Trade: The Hidden History of Prostitution Unveiled*, Lochlainn Seabrook.
- *Christmas Before Christianity: How the Birthday of the "Sun" Became the Birthday of the "Son"*, Lochlainn Seabrook.
- *Jesus and the Gospel of Q: Christ's Pre-Christian Teachings as Recorded in the New Testament*, by Lochlainn Seabrook.
- *Jesus and the Law of Attraction: The Bible-Based Guide to Creating Perfect Health, Wealth, and Happiness Following Christ's Simple Formula*, by Lochlainn Seabrook.
- *The Bible and the Law of Attraction: 99 Teachings of Jesus, the Apostles, and the Prophets*, by Lochlainn Seabrook.
- *Christ Is All and In All: Rediscovering Your Divine Nature and the Kingdom Within*, by Lochlainn Seabrook.
- *The Secret Jesus*, by Lochlainn Seabrook.
- *The Goddess in the Gospels: Reclaiming the Sacred Feminine*, Margaret Starbird.
- *Holy Blood, Holy Grail*, Michael Baigent, Richard Leigh, and Henry Lincoln.
- *Myth, Religion and Mother Right*, Johann Jakob Bachofen.
- *The Civilization of the Goddess: The World of Old Europe*, Marija Alseikait Gimbutas.
- *The Da Vinci Code*: Dan Brown.
- *The Jesus Mysteries: Was the Original Jesus a Pagan God?*, Timothy Freke and Peter Grandy.
- *The Mothers: The Matriarchal Theory of Social Origins*, Robert Stephen Briffault.
- *The Mystery-Religions and Christianity: A Study of the Religious Background of Early Christianity*, Samuel Angus.
- *The Isle of Avalon: Sacred Mysteries of Arthur and Glastonbury*, Nicholas R. Mann.
- *The Greek Myths*, Robert Graves.

- *The White Goddess: A Historical Grammar of Poetic Myth*, Robert Graves.
- *Goddesses' Mirror: Visions of the Divine From East and West*, David Kinsley.
- *The Book of Goddesses and Heroines*, Patricia Monaghan.
- *The Book of the Goddess, Past and Present: An Introduction to Her Religion*, Carl Olson (editor).
- *The Hebrew Goddess*, Raphael Patai.
- *Goddess Sites: Europe*, Anneli S. Rufus and Kristan Lawson.
- *When God Was a Woman*, Merlin Stone.
- *Sanctuaries of the Goddess: The Sacred Landscapes and Objects*, Peg Streep.

FALSE SCIENCE

- *Science Was Wrong: Startling Truths About Cures, Theories, and Inventions "They" Declared Impossible*, Stanton T. Friedman and Kathleen Marden.

UFO (AND RELATED) GROUPS, CLUBS, & ORGANIZATIONS

The second step to becoming a ufologist is to join a local, national, or international UFO group, club, or organization. There are hundreds of these to choose from, and there are new ones popping up every day.

Belonging to a UFO group will not only give you a chance to learn from others, but it will also give you a chance to share the knowledge you've learned as well. You'll make new friends and find lots of support for your interest in unidentified flying objects.

As a member of a UFO group you'll have the opportunity to attend meetings, lectures, seminars, and go on UFO watching field trips. Your group or organization will also keep you up to date on all of the latest UFO information, such as sightings, documents, and books.

Below is an international list of UFO groups, clubs, and organizations. There are many thousands of these. The following partial list is only meant to help you get started.

ARGENTINA

- FAO (Fundación Argentina De Ovnilogía): www.glaucoart.com.ar

AUSTRALIA

- ACERN (Australian Close Encounters Resource Network): www.acern.com.au
- ACUFOS (Australian Centre for UFO Studies): www.acufos.asn.au
- AUFORN (Australian UFO Research Network): www.auforn.com
- TUFOIC (Tasmanian UFO Investigation Centre): www.tufoic.ne1.net
- UFO & Paranormal Research Society of Australia: www.ufosociety.net.au
- UFOESA (UFO Experience Support Association): www.ufoesa.com
- UFOIC (UFO Investigation Centre): www.theozfiles.com
- UFO Research (NSW): www.ufor.asn.au
- UFO Research Queensland: www.uforq.asn.au
- UFOSA (Unhuman Flying Objects South Australia): http://ufosa.wordpress.com
- VUFORS (Victorian UFO Research Society): http://members.ozemail.com.au/~vufors

BELGIUM

- COBEPS (Le Comité Belge pour l'Étude des Phénomènes Spatiaux): www.cobeps.org
- SOS OVNI Belgique: www.sosovnibelgique.com

BRAZIL

- BURN (Brazilian UFO Research Network): www.burn.com.br
- CBPDV (Centro Brasileiro de Pesquisa de Discos Voadores): www.ufo.com.br
- CUB (Centro de Ufologia Brasileiro): www.cubbrasil.net

CANADA

- ASEPI (Association Sciences de l'Etrange and Phenomenes Inexpliques): www.asepiinc.org
- Quebec UFO Research: http://quebec-ufo-research.com

BULGARIA

- FOCONI (Foundation for Cosmonoetic Investigations): www.ufonews.in

CHILE

- CIO (Corporacion para la Investigacion Ovni de Chile): www.cio.cl

CHINA

- Hong Kong UFO Club: www.ufo.org.hk

CZECH REPUBLIC

- Project Zare: www.ufo.cz/zare

DENMARK

- SUFOI (Skandinavisk UFO Information): www.sufoi.dk

FINLAND

- UFO-Finland: www.ufofinland.org

FRANCE

- CERPA (Centre d'Etudes et de Recherches sur les Phénomènes Aérospatiaux): www.ufos-uhn.org
- Les Repas Ufologiques: www.les-repas-ufologiques.com

GERMANY

- CENAP (Centrales Erforschungs-Netz außergewöhnlicher Himmels-Phänomene): www.alien.de/cenap
- Independent Alien Network: www.greyhunter.de

INDONESIA

- Beta-UFO Indonesia: www.betaufo.org

IRELAND

- UFO Research Association of Ireland: www.ufoi.org
- UFO Society of Ireland: www.ufosocietyireland.com

ITALY

- CISU (Centro Italiano Studi Ufologici) : www.cisu.org
- CUN (Centro Ufologico Nazionale): www.cun-italia.net

LATVIA

- UFOlats: www.ufo.lv

MALAYSIA

- Malaysian UFO Network: www.myufo.net

MEXICO

- ALCIONE (Centro de Estudios Paranormales y Aeroespaciales Anómalos de México): www.alcione.org
- CONAIPO (Centro Internacional de Estudios Espaciales): www.conaipo.org
- La Esfera Azul: www.laesferaazul.tk

THE NETHERLANDS

- Skywatch Groningen: www.skywatch.nl.
- UFO Werkgroep Nederland: www.xs4all.nl/~ufonet

NEW ZEALAND

- UFO Focus New Zealand Research Network: www.ufocusnz.org.nz

NORWAY

- UFO-Rogaland: www.ufo-rogaland.no

POLAND

- Fundacja Nautilus: www.nautilus.org.pl
- INFRA: www.infra.org.pl

PORTUGAL

- Associação Pesquisa Ovni: www.apovni.org
- Sociedade Portuguesa de Ovnilogia: www.spo-ovnilogia.com

PUERTO RICO

- Puerto Rico UFO Network: www.prufon.com

ROMANIA

- Romanian UFO Network: www.rufon.org

SCOTLAND

- East 2 West UFO Society: www.e2wufos.org.uk

SPAIN

- CIFE (Centro Investigador Fenómenos Extraños: http://members.tripod.com/~cife
- Fundación Anomalía: www.anomalia.org: http://sib.org.es

SWEDEN

- UFO-Sweden: www.ufo.se/english

SWITZERLAND

- GREPI (Groupe de Recherche et d'Etude des Phenomenes Insolites): www.ovni.ch

TURKEY

- Sirius UFO Space Sciences Research Center: www.siriusufo.org

UNITED KINGDOM

- Birmingham UFO Group: www.bufog.com
- BUFOS (Bolton UFO Society): www.boltonufosociety.piczo.com

- BUFORA (British UFO Research Association): www.bufora.org.uk
- Contact International UFO Research: http://contactinternationalufo.homestead.com
- Conwy UFO Group: www.conwyufogroup.piczo.com
- Cornwall UFO Research Group: www.cornwall-ufo.co.uk
- Cosmic Conspiracies: www.cosmic-conspiracies.com
- Hull UFO Society: www.hufos.karoo.net/homepage.html
- LAPIS (Lancashire Anomalous Phenomena Investigation Society): www.lapisufo.com
- LUFOS (London UFO Studies): www.crowdedskies.com
- Southend UFO Group: www.essexufo.co.uk
- SUFOR (Swindon UFO Research): www.sufor.org.uk
- Truth Seekers Midlands: http://truthseekersmidlands.tripod.com
- Unknown Phenomena Investigation Association: www.upia.co.uk

UNITED STATES OF AMERICA

- CUFOS (Center for UFO Studies): www.cufos.org/index.html
- Fund for UFO Research: www.fufor.com
- llinois Mutual UFO Network: www.illinoismufon.com
- IRAAP (Independent Researchers' Association for Anomalous Phenomena): www.iraap.org
- ICAR (International Center For Abduction Research): www.ufoabduction.com
- International UFO Congress: www.ufocongress.com
- Kentucky MUFON: www.kymufon.org
- Michigan MUFON: www.mimufon.org
- MAARS (Missouri Awareness and Alternative Research Society): www.maarsgroup.org
- MUFON (Mutual UFO Network): www.mufon.com
- Museum of the Unexplained: http://ufoevidence.conforums.com/index.cgi
- NARCAP (National Aviation Reporting Center on Anomalous Phenomena): www.narcap.org
- National UFO Reporting Center: www.ufocenter.com

- NEUFOR (New England UFO Research Organization): www.neufor.com
- OPUS (Organization for Paranormal Understanding and Support): www.opus-net.org
- Organization for SETV Research: www.setv.org
- PAAPSI (Paranormal and Alien Abduction Problem Solvers International): www.delusionresistance.org/AACCOA
- Roundtown UFO Society: http://roundtownufosociety.com
- Shadow Research, Inc.: www.shadowresearch.com
- Tennessee MUFON: www.tnmufon.com
- UFORCE (UFO Resource Center): www.uforc.com
- UFO Wisconsin: www.ufowisconsin.com
- UFOzarks: http://UFOzarks.org
- Zerotime UFO Research: www.zerotime.com/ufo

CROP CIRCLES, VORTEX CENTERS,
& RELATED PHENOMENA

- CCCS (Centre for Crop Circle Studies): SKS, 20 Paul Street, Frome, Somerset, BA11 1DX, UK.
- CERES (Circles Effect Research Society): 3 Selborne Court, Tavistock Close, Romsey, Hants S051 7TY, UK.
- CNACCS (Center for North American Crop Circle Studies): P.O. Box 4766, Lutherville, MD 21904, USA.
- CPH (Circles Phenomenon Research): P.O. Box 3378, Branford, CT 06405, USA.
- CR (Circles Research): 1 Louvain Road, Greenhithe, Kent, DA9 9DY, UK.
- PRA (Phenomenon Research Association): 12 Tiltion Grove, Kirk Hallam, Ilkeston, Derbyshire, DE7 4GR, UK.
- NAICCR (North American Institute for Crop Circle Research): 649 Silverstone Avenue, Winnipeg, Manitoba R3T 2V8, Canada.
- VS (Vortex Society): P.O. Box 948, Sedona, AZ 86339, USA.

ALIEN ABDUCTION

- IF (Intruders Foundation): P.O. Box 30233, New York, NY 10011, USA.
- IFUFOCS (Institute for UFO Contactee Studies): 1425 Steele Street, Laramie, WY 82070, USA.
- John E. Mack Institute: www.johnemackinstitute.org
- MUFON (Mutual UFO Network): www.mufon.com
- TREAT (Center for Treatment and Research of Experienced Anomalous Trauma): P.O. Box 728, Ardsley, NY 10502-0728, USA.
- UFOCCI (UFO Contact Center International): 3001 South 288th Street, Suite 304, Federal Way, WA 98003, USA.

UFO & ALIEN DVD TITLES

- Aliens: the Complete Truth
- Bob Lazar, the Man Who Worked on a UFO
- Alien Implants
- UFOs: Greatest Story Ever Denied
- Alien Technology
- Bill Hamilton Blows the Lid on Alien Military Bases
- Area 51: Secrets of Dreamland
- Are We Alone in the Universe?
- Chariots of the Gods
- The NASA-Cydonia Briefings
- Close Encounters: Proof of Alien Contact
- Communion
- Crop Circles and Signs
- Alex Collier: Andromedan Update
- English Sacred Sites: The Atlantis Connection
- Evidence: The Case for NASA UFOs
- Flying Saucers Are Real
- God, Man, and ET
- Mars Revealed
- Hoagland's Mars
- Kecksburg: The Untold Story
- Left at East Gate

- Oz Encounters: UFOs in Australia
- Remote Viewing Methods: ESP From the Inside Out
- Secret NASA Transmissions: The Raw Footage
- Adventures Beyond: Chupacabra
- The Secret KGB UFO Files
- The Outer Space Connection
- The Scientific Study of Alien Implants
- UFO and Paranormal Phenomena
- UFO: They're Here!
- Evidence: The Case fo NASA UFOs
- UFOs Are Real
- UFO Abductions
- UFO Chronicles
- UFO Insiders
- UFOs: The Best Evidence
- UFO: The Best UFO Sightings of the 1990s
- UFOs: 50 Years of Denial
- UFOs: The Footage Archives
- UFO Contactee Ernie Sears
- UFOs: The Secret Evidence
- UFOs: The Hard Evidence (volumes 1-7)
- UFOs and Cosmic Dimension
- UFOs Uncensored: The Secret UFO Files and Abduction Files
- UFOs Over Phoenix
- UFOs and Area 51: The Bob Lazar DVD
- UFOs: Friend, Fantasy, or Foe?
- Ultimate UFO: The Complete Evidence
- Unsolved Mysteries: UFOs
- Hard Evidence: 3 Top UFO Documentaries
- Visitors: California UFO Wave
- What Happened on the Moon?
- Fastwalkers: Open Files
- UFO Hunters: Hoax or History? (Season One)

UFO & ALIEN-RELATED DVD TITLES

- Mind Control: For Reasons of National Security
- ET Mentoring
- The Missing Secrets of Nikola Tesla
- The Destruction of Atlantis
- Owning the Weather: The Secret Agenda of Atmospheric Manipulation
- Converting Zero Point Energy to Electricity
- A Funny Thing Happened On the Way to the Moon
- The Forbidden Secret
- CIA Mind Control: Out of the Darkness, Into the Light
- Free Energy, Anti-Gravity and Falling Magnetic Motors
- Free Energy: The Race to Zero Point
- Invisible Empire
- Project Pegasus, Time Travel, Teleportation and the Chronovisor Device
- Mars, Pyramids and Changes in the Solar System
- The Ancient Astronauts Who Built the Pyramids
- Stargate in the Middle East - Exposed!
- Brotherhood of Darkness
- Secrets of the CIA
- The Sons of God
- Underworld
- Water: The Great Mystery
- The Naked Truth
- The History of Mind Control
- HAARP: The Ultimate Weapon of the Conspiracy
- Dreamland Whistleblower Steps Into the Light
- Revelations of a Mother Goddess
- The Empire of the City: World Superstate
- Healing the Luminous Body: The Way of the Shaman
- The True Purpose of Our DNA
- Physics, Metaphysics and the Consciousness Connection
- The Living Matrix
- Spirit Space
- Dragons or Dinosaurs?

- Anunnaki - Coming Back?
- The Truth About the Ark of the Covenant
- Lost Secrets of the Sacred Ark
- Toxic Religion
- Ceremonial Magic
- Constitution Class
- Reptiles and PGLFs
- The Secret Underground Lectures of Commander X
- Ancient Code
- Bob Dean Exposes Life on the Moon and Mars
- The Creation of the Illuminati Bluebloods
- The Shadow Government
- Classified Projects Exposed!
- Behold A Pale Horse
- Hidden Symbols
- Dead Doctors Don't Lie
- Mystery of the Crystal Skulls
- Life After Life
- The Hidden History of the Human Race
- Man's True Genetic Origin
- Human Slave Species for ET Gods
- Celestials and ET Contact
- Contact Has Begun
- Structures on Saturn
- Moonviews
- Underground Tunnels and ET Technology

UFO NEWS CLIPPING SERVICES

Serious ufologists, especially ufologists who are writers, buy subscriptions to what are called UFO news clipping services. For a small subscription fee these services will photocopy and mail or email you articles from newspapers and magazines from all over the world every month.

Here's a list of a few UFO news clipping services:

- British UFO Newsclipping Service: CETI Publications, 247 High Street, Beckenham, Kent, BR3 1AB, England.
- UFO Newsclipping Service: #2 Caney Valley Drive, Plumerville, AR 72127-8725, USA.
- Worldwide UFO Newsclipping Bureau and Public Information Center: 955 West Lancaster Road, Suite 420, Orlando, FL 32809, USA.

UFOS ONLINE: WEBSITES PERTAINING TO UFOS

If you have computer, and you have access to the Internet, you'll want to investigate the many UFO Websites and UFO newsgroups that are now online. These range from the serious and the scientific to the humorous and the skeptical. While some require you to become a member and pay a fee before logging in, most are free and open to all.

An advantage of UFO Websites is that you can explore them any time day or night. They're also packed with current information, as well as incredible photographs of UFOs and aliens, all which are at your fingertips instantly.

To find the latest Websites dedicated to UFOs, simply type the words "UFOs" or "aliens" into your favorite search engine.

UFO MAGAZINES & JOURNALS

Another good source of information for ufologists are the many UFO magazines and journals that are available.

Some of these come out monthly, some bimonthly, some quarterly. You can subscribe to them, or buy them at your local bookstore or magazine shop each month as they appear; some are also available online. Most offer old back issues for sale, so you can study up on past UFO events.

Loaded with current news on the UFO phenomenon, UFO magazines also include interviews with famous ufologists and military people, informative articles, mind-bending photographs and stills from videos of UFOs, letters from readers, UFO book reviews, lists of upcoming UFO conferences, and even UFO classified sections.

Here's a list of UFO magazines and journals for you to choose from:

- *Communique*: P.O. Box 382 Woden, ACT 2606, Australia.
- *Continuum*: P.O. Box 172, Wheat Ridge, CO 80034-0172, USA.
- *Fate*: 84 South Wabasha Street, St. Paul, MN 55107, USA.
- *Flying Saucer Digest*: P.O. Box 347032, Cleveland, OH 44134-7032, USA.
- *Flying Saucer Review*: P.O. Box 162, High Wycombe, Buckinghamshire, HP13 5DZ, UK.
- *The International UFO Library Magazine*: 11684 Ventura Boulevard, Suite 708, Studio City, CA 91604, USA.
- *International UFO Reporter*: 2457, W. Peterson Avenue, Chicago, IL 60659, USA.
- *Journal of UFO Studies*: The J. Allen Center for UFO Studies, 2457 W. Peterson Avenue, Chicago, IL 60659, USA.
- *MUFON UFO Journal*: P.O. Box 369, Morrison, CO., 80465-0369, USA. Website: www.mufon.com
- *Northern UFO News*: 37 Heathbank Road, Cheadle Heath, Stockport, Cheshire, SK3 0UP, UK.
- *Strange*: P.O. Box 2246 Rockville, MD 20847, USA.
- *UFO*: P.O. Box 1053, Sunland, CA 91041-1053, USA.
- *UFO Encounters*: P.O. Box 1142, Norcross, GA 30091, USA.
- *The UFO Enigma*: P.O. Box 31544, St. Louis, MO 63131, USA.
- *The UFOlogist*: P.O. Box 1359, Palatka, FL 32178, USA.
- *UFO Magazine*: 27 A Clinton Street, Lambertville, NJ 08530, USA. Website: www.ufomag.com
- *UFO Magazine*: Wharfebank House, Wharfebank Business Centre, Ilkley Road, Otley, West Yorks, LS21, 3JP, England. Website: www.ufomag.co.uk
- *UFO Report*: P.O. Box 1144, Marshfield, WI 54449, USA.
- *UFO Reporter*: P.O. Box Q95, Queen Victoria Building, Sydney, New South Wales 2000, Australia.

UFO AMATEUR RADIO STATIONS

If you own an amateur radio, or have access to one, there are a number of UFO amateur radio stations you can tune into.

Here's a few in the US:

- ARUFON (Amateur Radio UFO Net), USA

Frequency: 3.9777 MHZ
Days: Tuesdays and Saturdays
Time: 01:00 Zulu (7:00pm CST to 8:00pm EST)
Mode: Lower Side Band (LSB)
Web: http://arufon.org

- MUFON Amateur Radio Net, USA

Frequency: 3.9777 MHZ
Day: Wednesday
Time: 01:00 Zulu (7:00pm CST to 8:00pm EST)

UFO CONFERENCES

Ufologists attend UFO conferences whenever they can. These get-togethers can last for an afternoon, or they can span several days. Usually noted ufologists, scientists, military personnel, and government employees are scheduled to speak or give lectures.

At a UFO conference you'll meet many new people, hear fascinating speakers, and learn about the latest theories and concepts behind the UFO phenomenon.

For information on times, dates, prices, and the scheduled guests of forthcoming UFO conferences, check with your local UFO group, a UFO Website, or any UFO magazine.

Here's a list of groups who hold conferences:

- Gulf Breeze UFO Conference: Project Awareness, P.O. Box 730, Gulf Breeze, FL 32562, USA.
- International MUFON Symposium: P.O. Box 369, Morrison, CO., 80465-0369, USA. Website: www.mufon.com
- International UFO Congress: 4266 Broadway, Oakland, CA 94611, USA.
- National UFO Conference: UAPA (United Aerial Phenomenon Agency): P.O. Box 347032, Cleveland, OH 44134-7032, USA.
- NICUFO (National Investigations Committee on UFOs) Annual Conference: P.O. Box 73, Van Nuys, CA 91408-0073, USA.

- Rocky Mountain UFO Conference: IFUFOCS (Institute for UFO Contactee Studies), 1425 Steele Street, Laramie, WY 82070, USA.
- San Francisco Expo West: P.O. Box 1011, Pacific Palisades, CA 90272, USA.
- The UFO Experience Annual Conference: P.O. Box 2051, Cheshire, CT 06410, USA.

STUDY THE GLOSSARY

Finally, a good ufologist is an educated ufologist. This means having a keen grasp of the words, acronyms, abbreviations, and phrases that are often used in ufology.

Throughout this book words that you'll need to be familiar with appear in bold print. For example: **unidentified flying object**. This tells you that this phrase is in the glossary, where you'll find a full definition or explanation of what it means.

Study the glossary. Try using these words and phrases in your daily conversations and in your notes and writings. Ask your family members or friends to test you on your knowledge.

You are now on your way to becoming a real ufologist!

CHAPTER 12 AT A GLANCE

- To become a ufologist, you must study, think, and learn.
- All ufologists have a sense of uforia.
- The first step to becoming a ufologist is to read every book on UFOs that you can.
- As you read, write down what you learn in your notebook.
- Join a UFO group, club, or organization.
- In a UFO club you'll meet lots of other people who share your interest in ufology.
- In a UFO club you can both learn from others and share your knowledge.
- In a UFO club you'll be able to go on UFO watching field trips and attend UFO meetings and conferences.

- In a UFO club you keep up with the latest UFO news.
- Buying a subscription to a UFO news clipping service is a good way for a ufologist to keep tabs on UFO sightings all over the world.
- Online UFO Websites offer a ufologist unlimited access to UFO information twenty-four hours a day, seven days a week.
- UFO magazines and UFO journals are a handy source of UFO information.
- Many wonderful TV programs on UFOs have been created, some from as early as the 1950s. All help add to our knowledge of the topic. Some of the better ones: *Chasing UFOs*, *UFO Hunters*, *UFO Files*, *UFOs Over Earth*, *Ancient Aliens*, *Project U.F.O.*, *One Step Beyond*, *In Search Of . . .*, and *Sightings*.
- Many ufologists tune into UFO amateur radio stations to hear discussions and news about UFOs.
- Attending UFO conferences will give you a chance to hear famous authorities on UFOs and aliens.
- Educate yourself. Read and study the glossary at the end of this book so that you can become familiar with words that are commonly used in ufology.

You now know everything you need to know to begin your own investigation into UFOs and aliens. Time to go outside and start looking!

Appendix A

UFO SIGHTING REPORT FORM

(Use back of sheet if necessary)

Date today:
My name:
My mailing or street address:
My email address:
My telephone number:
Date of sighting:
My location at the time of sighting:
How many objects:
The location of the object(s):
The speed of the object(s):
The shape of the object(s):
The size of the object(s):
The color(s) of the object(s):
The behavior of the object(s):
The direction the object(s) was moving:
Noises made by the object(s):
Lights on the object(s):
The altitude of the object(s):
Weather at the time of sighting:
Time of the sighting: (AM/PM)
Were the Sun, the Moon, or any stars visible?
Duration of the sighting: hours mins seconds
Did the object(s) pass in front of or behind anything?:
Were any photos, films, or videos made of the object(s)?:
Similarities to any known natural phenomena or human-made aircraft:

Other phenomena that occurred during sighting:

Other witnesses:

Describe the object(s) in detail:

Draw a sketch of the object(s):

What do you believe you saw? (circle one): star, planet, human-made object, jet, plane, helicopter, balloon, hoax, satellite, UFO, other:

Remarks (describe your emotions and any sensations you felt at the time of the sighting):

My signature:

Appendix B

SAMPLE UFO SIGHTING REPORT FORM

Date today: January 2, 2015.

My name: Erin O'Malley.

My address: 1234 Robert E. Lee St., My Town, Tennessee, 12345, USA.

My email address: erin@ufosandaliens.net

My telephone number: 555-555-1234.

Date of sighting: January 1, 2015.

My location at the time of sighting: on the street in front of my house.

How many objects: 1.

The location of the object(s): right above our house.

The speed of the object(s): from stationary to over 1,000 mph.

The shape of the object(s): cylinder- or cigar-shaped.

The size of the object(s): 50 ft. long, 15 ft. high, the size of a bus.

The color(s) of the object(s): silvery gray on top, bright blue on bottom, orange in middle.

The behavior of the object(s): it hovered, then descended toward the ground, then hovered and wobbled, then shot straight up and disappeared.

The direction the object(s) was moving: north, then north-west.

Noises made by the object(s): beeping, then hissing, after that it was completely silent.

Lights on the object(s): a blue light at front on bottom, a large amber light on top in the center, and a red light on rear bottom; all were flashing wildly and not in unison.

The altitude of the object(s): 500 ft., moved down to 10 ft.

Weather at the time of sighting: clear, cool, 30 degrees, no wind.

Time of the sighting: about 6:30 PM.

Were the Sun, the Moon, or any stars visible? no Moon, but some stars.
Did the object(s) pass in front of or behind anything?: yes, a cloud.
Duration of the sighting: hours 12 mins 32 seconds.
Were any photos, films, or videos made of the object(s)?: yes, 5 photos.
Similarities to any known natural phenomena or human-made aircraft: looked kind of like an airplane without wings or a tail section.
Other phenomena that occurred during sighting: saw other small bright flashing lights in the distance, toward the north.
Other witnesses: Brianna Kelly Flannigan. Tel: 555-555-5555. Address: 5555 Jefferson Davis St., My Town, Tennessee, 12345, USA.
Describe the object(s) in detail: it was a bizarre-looking thing, unearthly. It hovered in mid-air without wings and without making a sound! It had little windows. You could see lights on the inside. The colors of the body and the lights weren't like any plane or anything. It was too far away to see if it had bolts or rivets, or what to see it was made of, but on the outside it looked like an aluminum pan.
Draw a sketch of the object(s): enclosed.
What do you believe that you saw? (circle or underline): star, planet, human-made object, jet, plane, helicopter, balloon, hoax, satellite, <u>UFO</u>, other (describe):
Remarks (describe your emotions and any sensations you felt at the time of the sighting): I was amazed by what I saw. It was a beautiful object, but scary at the same time. The way it moved, and the colors and lights. It was like nothing I've ever seen before. My friend and I are both still in shock.

My signature: *Erin O'Malley*

GLOSSARY

CCOMMON WORDS, ABBREVIATIONS, ACRONYMS, AND PHRASES USED IN UFOLOGY

"[The Airship] looked like a great black cigar with a fishlike tail. . . . The body was at least 100 feet long and attached to it was a triangular tail, one apex being attached to the main body. The surface of the airship looked as if it were made of aluminum, which exposure to wind and weather had turned dark. . . . The airship went at tremendous speed. As it neared Lorin [California] it turned quickly and disappeared in the direction of San Francisco. At half past 8 we saw it again, when it took about the same direction and disappeared." — Case Gilson, Oakland, California, 1896

AAC: Alaskan Air Command.

AAF: Army Air Force.

abductee: an individual who has been kidnaped by extraterrestrials.

abductee implant: see "implant."

abduction: the act of an alien kidnaping a human or animal.

ADC: Aerospace Defense Command.

AEC: Atomic Energy Commission.

aerobatics: stunning aerial maneuvers.

aerolite: a stony meteorite.

AFB: Air Force Base.

AFGWC: Air Force Global Weather Control.

AFIS: Air Force Intelligence Service.

AFOC: Air Force Operations Center.

AFOSI: an acronym for America's "Air Force Office of Special Investigation."

AFR: Air Force Regulation.

AFSS: Air Force Security Services.

Age of Flying Saucers: though UFOs have been seen as far back as prehistoric times, the modern age of UFOs officially began in 1947, with the Roswell Incident.

aircraft: a man-made vehicle capable of flight, typically operating using wings and/or engines.

ADC: Air Defense Command.

airship: a lighter-than air vehicle capable of flight.

alien: a living being not from Earth; also known as an ET.

alien agenda: the aliens have a specific reason, or agenda, for visiting Earth. They seem to be interested in harvesting human DNA in an effort to make a hybrid species that's half alien and half human, one that can survive on our planet and carry on the alien genetic code. They also seem to be trying to save our planet from human pollution and nuclear war.

AMB: ambassador.

amnesia: a loss of memory, or a gap in memory, usually caused by shock or illness.

amnesiac block: a loss of memory, or a gap in memory, intentionally created to hide a painful or disturbing event or experience; it's been found that aliens usually block a person's memory of their abduction.

ANG: Air National Guard.

angel hair: a thin silvery material sometimes found on the ground after the appearance of a UFO; the substance evaporates within a few hours.

anomalies: something out of the ordinary.

APRO: Aerial Phenomena Research Organization.

ARPA: Advanced Research Projects Agency.

ASD: Applied Science Division.

AST: Atlantic Standard Time.

astrology: the science of how life is effected by the placement of the stars.

astronautics: the science of constructing and navigating space vehicles.

astronaut: a human who travels into space.

astrophysics: the scientific study of space matter.

ATIC: an acronym for America's "Air Technical Intelligence Center." On August 5, 1948, the ATIC released a document called an "Estimate of the Situation," which stated that it was their opinion that UFOs are real and that they "are interplanetary in origin." Government officials later ordered the document to be burned.

Atlantis: an ancient island located in the Atlantic Ocean, believed to have sunk beneath the sea due to some type of early natural catastrophe; it's inhabitants are said to have been of an advanced nature.

auras: an electrical or spiritual field that surrounds every living thing; some can see auras and even diagnose illnesses using them.

ball lightning: a rare type of lightning that forms into small sphere-like balls of bright light and gas. Scientists know literally nothing about the phenomenon.

black budget: a secret supply of money given to a government agency by the government to fund black projects or programs. Over the years black projects dedicated to UFOs have received billions of dollars.

black government: a secret government that operates within a larger government.

black project: a secret program created by a black government and funded by a black budget. Studying UFOs is only one of many types of black programs that the governments of the world are hiding.

blip: an image on a radar screen.

BMW: Bomb Wing.

bogey: military and NASA slang for a UFO.

CAA: in the U.S., Civil Aeronautics Administration; in Canada, the Canadian Aviation Administration.

CAB: Civil Aeronautics Board.

canopy: the window over an aircraft's cockpit.

CAUS: Citizens Against UFO Secrecy.

cereology: the scientific study of crop circles.

cereologists: experts who study crop circles.

CFS: Canadian Forces Station.

CGS: Coast Guard Station.

chthonic: meaning "earthly," one of two categories of deities in Greek mythology; the other category is Ouranian ("heavenly"); sometimes used to describe aliens.

CIA: an acronym for "Central Intelligence Agency," a branch of America's intelligence community. Formed in 1947, the CIA has a yearly budget of 1 billion dollars, and may have as many as 25,000 employees. U.S. government documents show that the CIA, like many other official agencies, has been involved in UFO research—and the suppression of UFO information—for many decades. Inside sources say that at the CIA headquarters in Langley, Virginia, there are an estimated 10,000 pages of classified UFO documents. The CIA says that it only has fifty-seven documents!

CIC: Counter-Intelligence Corps.

cigar-shaped: a UFO in the form of a cigar. Very common.

CINC: Commander-in-Charge.

cipher: secret coded communication.

clairaudient: an individual who can hear non-physical sounds.

clairvoyant: an individual who can see and feel non-physical things.

classified: something kept from the general public for purposes of national security.

cluster UFO: a group of independent unidentified flying objects that appear to be touching, or are touching, as they fly or hover. Recently a cluster UFO was seen by multiple eyewitnesses in South Korea that contained dozens of brilliant white silvery objects.

CNES: an acronym for the "Centre Nationale d'Études Spatiales," France's "NASA."

CNRS: an acronym for France's "Center for Scientific Research."

COMINT: Communications Intelligence.

COMSEC: an acronym for "Communications Security," a branch of Britain's intelligence community.

contactee: a person who's been abducted by aliens. The same as "abductee."

contactology: the science of studying the interaction between humans and aliens (usually in relation to abductions).

contrail: short for "condensation trail," a contrail is a cloud-like trail of frozen water vapor (ice particles) that's left behind jets and also sometimes UFOs.

conventional aircraft: everyday well-recognized, human-made aircraft, like planes, gliders, jets, etc.

cosmic consciousness: a level of consciousness beyond the ordinary; an advanced state of mind. All of the great spiritual teachers are said to have attained it (Jesus, Buddha, Krishna, Zoroaster, etc.)

cosmonauts: the Russian word for astronaut.

CP: Command Post.

cryptography: the science of secret or coded writing.

cryptozoology: the scientific study of unknown animals; closely connected to ufology.

CSC: Central Security Control.

CUFOS: an acronym for the American based organization, "Center for UFO Studies." Today CUFOS is based in Chicago, IL, and is sometimes referred to as the "J. Allen Hynek Center for UFO Studies."

cult group: a group, usually religiously or spiritually oriented, that exists outside the mainstream; often anti-social and not therefore accepted by society (though not always).

cusp: on the edge of.

DATT: Defense Attache.

DCD: Domestic Collections Division.

DCSOPS: Deputy Chief of Staff for Operations and Plans.

DD: in ufology these initials stand for "daytime disk," or "diurnal disk." DD refers to a UFO that's been sighted during daylight hours.

DDO: Deputy Director for Operations.

debunk: to disprove or expose an idea or belief that's thought to be false.

debunker: a person who tries to disprove an idea or belief that they feel is false.

delta-winged: an aircraft with triangular-shaped wings or body.

demonism: the study of demons; also the practice of worshiping demons.

demonology: the scientific study of demons.

DIA: an acronym for the "Defense Intelligence Agency," a branch of America's intelligence community. U.S. government documents show that the DIA, like many other official agencies, has been involved in UFO research for many decades.

DIALL: an acronym for Britain's "Defense Intelligence Agency Liaison in London." DIALL is part of the MOD.

DIS: an acronym for "Defense Intelligence Staff," the DIS is a branch of Britain's Ministry of Defense.

discs (or **disks**): disk- or saucer-shaped UFOs.

disinformation: intentionally leaking false information in an effort to confuse, deceive, and mislead. As an example, the U.S. and British governments spread disinformation about UFOs and aliens to prevent the public from knowing the truth. They also plant fake stories in newspapers and magazines and put out phony, and often harmful, information that discredits UFO eyewitnesses.

DMZ: Demilitarized Zone.

DO: Duty Officer.

dogfight: an air-battle between flying craft.

DST: an acronym for "Direction de la Surveillance du Territoire," a branch of France's intelligence community. The DST is the same as America's FBI or Britain's MI5.

Earth: our planet, named after the old Pagan European Earth-Goddess Erda. Earth is only one of possibly trillions of planets in the known Universe.

earthian: related to the Earth.

earthlight: reflected sunlight cast by the Earth back into space.

earthling: a being from Earth.

EBE: Extraterrestrial Biological Entity, the scientific term for an alien.

echelon: a formation of aircraft in flight wherein each vehicle is positioned at a different altitude and elevation to the one ahead of it.

ECM: an American acronym for "electronic countermeasures equipment."

ELINT: an American acronym for "electronic intelligence." ELINT is information gathered by machines rather by humans (see HUMINT).

EM: electromagnetics.

EME: an acronym for "electromagnetic energy." The term refers to the strange ability of UFOs to interfere with human-made electronic devices, which begin to malfunction in the presence of UFOs. These would include, cars, jets, cameras, lights, TVs, compasses, cell phones, radios, etc.

ET: an acronym for "ExtraTerrestrial" (terrestrial means "of the Earth," and extra means "beyond"; thus, "beyond Earth"). An ET is any living creature that's not from Earth, even a bacterium.

ETH: an acronym for the "extraterrestrial hypothesis," the theory or belief that UFOs exist and that they're made and controlled by an alien civilization.

extrasensory perception: ESP is the ability to sense things beyond the physical.

FAA: an acronym for America's "Federal Aviation Administration."

FBI: an acronym for the "Federal Bureau of Investigation," a branch of America's intelligence community. France's "FBI" is called the DST, while Britain calls theirs MI5.

fireball: any fiery object seen in the heavens, from a comet or meteor to a UFO or crashing plane.

flak: exploding shells from antiaircraft guns.

flap: a state of panic.

flying saucer: a term coined in the 1940s; used to describe one of hundreds of types of UFOs.

flying wing: an aircraft design in the shape of a giant single wing; notable past examples include the Northrop XB-35, and today's Northrop Grumman B-2 "Spirit"; UFOs also appear in the flying wing design, and have for many centuries.

FOIA: an acronym for the "Freedom of Information Act." In the U.S. the FOIA began in 1966 and was later reinforced in 1974. The FOIA gives an individual the right to request information from government agencies, who must respond within ten business days. Some information—such as highly classified military, law enforcement, and intelligence files—cannot be obtained through the FOIA. Ufologists have found that the government often uses these exemptions as an excuse to hide many important documents on UFOs and aliens, but which the public has a right to see. Britain recently instigated their own FOIA, and the act was passed in Canada in 1982.

foo fighter: strange balls of light (about five feet in diameter) that accompanied pilots during flight in World War II, the Korean War, and in Vietnam; each side assumed it was an enemy vehicle, but foo fighters easily outmaneuvered any manmade aircraft of that day.

formation: the flight pattern of multiple aircraft.

fuselage: the body of an aircraft.

galaxy: a large system comprised of stars, solar systems, nebulae, and interstellar matter; the Universe contains countless billions of galaxies.

GAO: an acronym for the "General Accounting Office," a "watchdog" department of the U.S. government that keeps tabs on how the government spends taxpayers' money. It has recently been renamed the "Government Accountability Office," though the acronym remains the same.

GCHQ: an acronym for "Government Communications Headquarters," a branch of Britain's intelligence community.

GCI: Ground Control Intercept.

GEPAN: an acronym for the "Groupe d'Études Phénomènes Aerospatiaux Non Identifiés," a French government agency set up in 1977 to study UFOs. After GEPAN found that UFOs are real and that they come from outside Earth, certain elements of the government's intelligence community tried to shut it down.

ghost rockets: an early name for UFOs in Scandinavia.

GMT: Greenwich Mean Time.

GSW: Ground Saucer Watch.

hard evidence: concrete proof.

hard sightings: irrefutable eyewitness sightings of UFOs and aliens.

high strangeness: a term coined by astronomer J. Allen Hynek to describe the unique strangeness of UFOs; a strangeness that's far beyond normal everyday strangeness. More specifically the term refers to the many baffling and even bizarre mysteries that surround the UFO phenomenon.

hoax: an intentional fraud or scam; in UFO terms, creating a fake UFO.

hostile: in aeronautic terms, an enemy vehicle, often of an aggressive nature.

hot spots: places where UFOs have crashed, or where they seem to congregate, and where UFO watchers have a better than average opportunity of seeing a UFO.

Military bases are one of the hottest hot spots. UFO hot spots also include places where odd military activity has taken place. UFO enthusiasts and skeptics alike travel from all over the world to UFO hot spots.

humanoid: human-like; having the characteristics of a human being.

HUMINT: an American acronym for "human intelligence." HUMINT is information gathered by humans rather than by machines (see ELINT).

hybrid: half and half; something created from two different sources.

hypnotic regression: hypnosis in which an individual is taken back in time.

identified flying object: see IFO.

IFO: an acronym for "Identified Flying Object." In a sense an IFO is the opposite of a UFO. Almost all IFOs start out being mistaken for UFOs. In fact, 90 percent of "UFOs" later turn out to be IFOs, which would include stars, clouds, planets, planes, helicopters, car headlights, hoaxes, etc.

implant: a small device, usually made of organic metal, that aliens surgically place in a person's body to keep track of them. An alien implant seems to be similar to the bar code that stores put on their products. They create no inflammatory response in the body and so often go undetected.

interdimensional: another word for paranormal.

intergalactic federation: an off-earth organization made up of different alien races, similar to the one represented on *Star Trek*.

interplanetarians: aliens who live on other planets.

INYSA: Assistant Chief of Staff, Intelligence.

INZ: Aerospace Intelligence Division.

INZA: Editing, Debriefing, and Continuity Branch.

JACL: Judge Advocate General, Litigation Division.

JANAP: an acronym for America's "Joint Army-Navy-Air Force Publication." On February 17, 1954, JANAP imposed severe restrictions on all military personnel concerning UFOs. According to JANAP 146, no military person is allowed to talk about their UFO sightings with either the public or the media. The penalty if caught is a $10,000 fine and ten years in prison.

JCS: Joint Chiefs of Staff.

jovian: related to the planet Jupiter; also an early name for God ("Jove"; i.e., the Roman Father-God Jupiter, who is identical to Zeus, Yahweh, and Jehovah; see Acts 17:22-23).

JPL: an acronym for NASA's "Jet Propulsion Laboratory." The JPL is in charge of NASA's unstaffed planetary space programs, such as the Viking and Voyager probes.

jupiterian: a being from Jupiter; or anything relating to Jupiter.

KGB: an acronym for Russia's Komitet Gosudarstvennoi Bezopastnosti ("Committee for State Security"). The KGB is the world's largest secret service agency, with over 2 million employees around the world. Like many secret service agencies, the KGB has a department especially devoted to creating coverups and spreading disinformation. Called "Department D," the KGB's office of disinformation has invented many false stories about UFOs over the years.

KISR: Kuwait Institute for Scientific Research.

KM: kilometer.

KSC: an acronym for America's "Kennedy Space Center," located at Cape Canaveral, Florida.

LCF: Launch Control Facility.

Lemuria: an island once located in the Southern Pacific Ocean or in the Indian Ocean; like Atlantis, its inhabitants were said to be of a highly spiritually evolved nature.

lenticular cloud: cloud formation that looks like a lens, and often mistaken for flying saucer-shaped UFOs.

ley line: mysterious invisible lines over Great Britain marking ancient sacred energy paths; sacred sites have been aligned over ley lines for thousands of years, revealing a large well-ordered grid.

LGM: little green men; or laser-guided missiles.

lunar: related to the Moon.

lunarian: a being from the Moon; anything related to the Moon.

magnetic: the power of attraction.

martian: a being from Mars; anything related to Mars.

mass sighting: a UFO sighting where there are large numbers of eyewitnesses. Some mass sightings include hundreds or thousands of people, while others, like those over Mexico City, include millions.

medium: one who channels the spirits of the dead.

mercurian: a being from Mercury; anything related to Mercury.

meteor: a piece of space matter that becomes temporarily incandescent when it enters Earth's atmosphere (due to friction); most vaporize before reaching the surface of the Earth.

meteorite: a meteor that survives vaporization and lands on the surface of the Earth.

meteoroid: a meteor that revolves around the Sun.

MI5: Britain's Security Service, identical to America's FBI and France's DST.

MI6: the Secret Intelligence Service, a branch of the British intelligence community. Also known as SIS, MI6 is Britain's equivalent to America's CIA.

MIB: an acronym for "men in black." The MIB are men in expensive black suits who look human but act strangely alien. They're said to visit and harass those who witness UFOs, or who study, write, or talk about UFOs.

millennium: one thousand years.

mirage: not an optical illusion as is commonly thought, but a real phenomenon caused by atmospheric refraction. There are numerous types of mirages: the mock mirage, the inferior mirage, and the three-image mirage, to name a few. Skeptics claim that many UFO sightings are nothing but mirages. However, no mirage flashes bright lights, hovers, makes ninety-degree turns, and travels at thousands of miles an hour.

missing time: a phrase referring to the fact that time seems to stop, or "stand still," when UFOs come into an area, or when a person is abducted by aliens. Some people report only a few seconds missing, others report hours, or even days, that can't be accounted for.

MJ-12: an acronym for Operation "Majestic-12," a code-name for a secret panel of twelve men who were chosen by American President Harry Truman in 1947 to study UFOs. According to top secret documents, the panel eventually made a deal with the aliens, which allowed them to abduct humans for medical research in exchange for the secret of UFO technology. The twelve men were: Dr. Lloyd Berkner, Dr. Detlev Bronk, Dr. Vannevar Bush, James Forrestal, Gordon Gray, Vice Admiral Roscoe Hillenkoetter, Dr. Jerome Hunsaker, Dr. Donald Menzel, General Robert Montague, Rear Admiral Sidney Souers, General Nathan Twining, and General Hoyt Vandenberg. All twelve men are now deceased.

mock suns: also called "sun dogs," or parhelia, mock suns are formed when sunlight is refracted by hexagonal plate-like ice crystals, which create a brilliant ball of light on either side of the rising or setting Sun; they're sometimes confused with UFOs.

MOD: an acronym for Britain's "Ministry of Defense."

Moon: the name of Earth's only satellite, but just one of 167 moons currently recognized by NASA in our solar system (Jupiter alone has sixty-two known moons). Our Moon reveals many irregularities, all or most seeming to indicate artificiality, and thus an alien presence. For example, our Moon is the only moon in our solar system that doesn't rotate on its own axis (as a result, we can't see the Moon's other side, known as the "dark side," from Earth), and which also has a nearly perfectly stationary circular orbit. Strangely, the Moon is also the exact diameter and the exact distance from Earth to fully cover the Sun, and thus create a solar eclipse (the odds of this occurring *naturally* are infinitesimal). Some studies seem to show that the Moon is hollow with a rigid metallic shell, and that it's actually many hundreds of thousands of years older than the Earth. Yet conventional astronomers say that the Moon was formed from the Earth! Still others believe that NASA photos show various artificially constructed objects on our Moon, such as towers, bridges, rectangles, pyramids, tubes, and domes (though these have been airbrushed out of most pictures by NASA photo technicians). Some photos seem to show that much of the surface of the Moon is covered with uniformly separated straight lines. Who made them, and why? Flashing and blinking lights (of all colors) have been observed by astronomers on our Moon for hundreds of years, as have weird mists, clouds, and shadows. NASA transmissions of astronauts near or on the Moon are said to contain references to UFO sightings. All of this, of course, has been carefully concealed from the public.

mother ship: large alien spacecraft used to transport smaller vehicles (UFOs) to and from distant regions of space; mother ships are often seen by abductees, and also by pilots during flight.

MTE: an acronym for "missing time experience." An MTE can occur during almost any type of close encounter with a UFO, but it's more likely to occur when a person has actual contact with aliens. Typically, after a UFO sighting or alien abduction, a person finds that they can't account for the amount of time that's passed. The missing time can be include minutes, hours, or even days.

MUFON: an acronym for America's "Mutual UFO Network." Headquartered in Colorado, MUFON's motto is: "Dedicated to the scientific study of UFOs for the benefit of humanity."

multiple abductee: a person who's been abducted by aliens more than once. Some multiple abductees have been taken dozens of times throughout their lives, beginning as early as infancy.

multiple sighting: any sighting of a UFO by more than one person.

mutologists: individuals who study animal mutilations.

mutology: the scientific study of animal mutilations.

mystery helicopter: an unmarked unidentified flying object that looks, acts, and sounds like a helicopter, but may not actually be a helicopter. It's generally believed that the mystery helicopter is either a black government craft or a UFO disguised to look like a helicopter.

NASA: an acronym for America's "National Aeronautics and Space Administration." Founded in 1958 by President Dwight Eisenhower, and headquartered in Washington, D.C., NASA is a nonmilitary government agency whose mission is space exploration; although many question this, citing numerous examples of how NASA has ignored, ridiculed, and suppressed potentially stunning new discoveries. NASA has a number of laboratories, research stations, and space flight launching centers, such as the Houston Control Center and Cape Canaveral. NASA has been secretly researching UFOs for decades, while denying that it has any interest in them, leading some to suggest that NASA is actually an acronym for "Never A Straight Answer." Many astronauts, in fact, have seen and filmed UFOs while in orbit, but NASA has confiscated these tapes and threatened the astronauts with fines and imprisonment to keep them from telling their stories to the public.

NAVSPASUR: an acronym for America's "Naval Space Surveillance System." NAVSPASUR observes the airspace above the U.S. for enemy aircraft and missiles. As government documents show, on many occasions NAVSPASUR has recorded genuine UFOs.

NCOC: an acronym for the "National Combat Operations Center."

NDE: an acronym for the "near-death experience."

NEPA: an acronym for the "Nuclear Energy for the Propulsion of Aircraft."

neptunian: a being from the planet Neptune; or anything related to Neptune.

NICAP: an acronym for the "National Investigations Committee on Aerial Phenomena." NICAP was founded in 1956 by Thomas Townsend Brown, a Navy physicist, and headed by Major Donald Keyhoe, a firm believer in UFOs. From its beginnings NICAP has been a staunch opponent of government secrecy, and little wonder. NICAP researched the topic of UFOs and found that they were real and that they come from outside our planet. After these findings were made, Keyhoe was silenced and finally removed, and NICAP was dismantled. In 1973, however, all of NICAP's files were taken over by CUFOS.

NL: in ufology these initials stand for "nighttime lights," or "nocturnal lights." NL refers to a UFO that's been sighted after dark, or at night.

NM: an acronym for a "nautical mile."

NMCC: an acronym for the "National Military Command Center."

NORAD: an acronym for America's "North American Aerospace Defense Command."

NPIC: an acronym for the "National Photographic Interpretation Center."

NRL: an acronym for the "Naval Research Laboratory."

NRO: an acronym for America's "National Reconnaissance Office." The NRO, set up in 1960, is the most secretive of America's secret intelligence agencies.

NSA: an acronym for America's "National Security Agency," set up in 1952. American government documents show that the NSA, like many other official agencies, has been involved in UFO research for many decades.

NSF: National Science Foundation.

OAM: Operation Animal Mutilation.

occult: belief and/or study in the paranormal.

ONI: Office of Naval Intelligence.

OOBE or **OBE**: out-of-body-experience.

optical illusion: a visually perceived image that's false, deceptive, or misleading. Skeptics like to pass off many UFOs sightings as optical illusions, but this is impossible in cases such as multiple sightings, mass sightings, and UFOs tracked on radar and sonar.

OSF: Objects Seen Floating.

OSI: Office of Special Investigations; or Office of Scientific Investigations.

OSS: an acronym for America's "Office of Strategic Services." The OSS investigated the "foo fighter" UFOs during World War II and found them "unusual but harmless." Not wanting the public to know the truth about UFOs, the OSS engaged in many disinformation projects, such as hoaxing UFO sightings and publishing anti-UFO stories in newspapers. The CIA was formed out of the OSS in

1947, at which time it adopted the OSS's "conspiracy of silence and denial."

ouranian: meaning "heavenly," one of two categories of deities in Greek mythology; the other category is chthonic ("earthly").

overflight: passing over an area in an aircraft.

paleoanthropologists: scientists who study prehistoric humans.

parachute canopy: the large umbrella-like cloth that opens up when a ripcord is pulled, allowing a pilot to land on the ground safely.

para-government: the same as a black government.

paranormal: beyond the norm; the supernatural.

phantasmagoria: objects that appear to quickly grow larger in size when moving toward the observer, and vice versa.

phenomenon: an observable fact, as perceived through the five physical senses.

PHOTINT: an American acronym for "photographic intelligence."

plasma: charged particles possessing an equal number of positive and negative ions and electrons.

plutonian: a being from the planet Pluto; or anything related to Pluto.

prehistory: the great expanse of time period before recorded history. There is much debate on just when history (written records) itself began; early writings appear all across Europe and the Near and Middle East, the earliest presently known being in Egypt, about 3,400 BCE; however, if we include art (as a form of record-keeping) in the definition of history, then prehistory ended about 500,000 years ago, when we find the first artistic images in the fossil record (probably created by *Homo erectus*).

project: a large governmental undertaking.

protoplasm: a complex of organic and inorganic materials that make up living tissue.

prototype: an original model, type, form, or pattern.

psychic: one who possesses extrasensory perception.

RAAF: an acronym for both the "Royal Australian Air Force" and the "Roswell Army Air Field."

radar: this is not actually a word, but rather an acronym, one that stands for "radio detecting and ranging." Radar is a radio device that sounds out ultra-high frequency radio waves which bounce off an object and are reflected back to the device. The information that the radar device receives is used to determine where the object is, how high or low it is, the direction it's moving, and hopefully what it is.

RADINT: an acronym for "radar intelligence," an aspect of Britain's intelligence community.

radioastronomy: a branch of astronomy that deals with receiving electromagnetic radiations of radio frequency from outer space. Radiation is so-called because it "radiates" (sends out rays of) energy in the form of waves or particles.

RAPCOM: Radar Approach Communications.

RAPCON: Radar Approach Control.

RAF: an acronym for Britain's "Royal Air Force."

RCMP: Royal Canadian Mounted Police.

Rosetta Stone: an ancient stone discovered by Captain Pierre-François Bouchard in Egypt in 1799; the stone contains an ancient text written in three languages: Egyptian demotic, Egyptian hieroglyphs, and Greek. Knowledge of Greek allowed scientists to decipher the mysterious hieroglyphs. Can also be a metaphor for something that's the key to understanding.

RPV: an acronym for "remotely piloted vehicle," also known as drones or probes. A satellite is an example of a human-made RPV. The aliens have their own: thousands of mysterious canister-like alien RPVs have been spotted hovering over various areas on Earth, patiently observing us.

RV: in ufology these initials stand for "radar visual." RV refers to a UFO that's been sighted, not by the human eye, but by the electronic eye called radar.

SA: special agent.

SAC: an acronym for America's "Strategic Air Command." There are many SAC bases around North America, such as in Canada, Montana, Maine, North Dakota, and Michigan. All of these have been overflown by UFOs; may also mean "special agent in charge."

SAC/HO: Strategic Air Command History Office.

SAFOI: Secretary of the Air Force Office of Information.

SAO: Smithsonian Astrophysical Observatory.

SAT: Security Alert Team.

saturnian: a being from the planet Saturn; or anything related to Saturn.

saucerian: an object shaped like a saucer.

saucerinism: the belief in and/or study of UFOs.

saucerite: a person who believes in UFOs; one who accepts UFO reality.

saucerological: anything related to UFOs.

saucerology: the scientific study of UFOs.

skeptic: when it comes to the paranormal, and in particular UFOs and aliens, a "Doubting Thomas."

scramble: to take off quickly in response to a warning or problem; to move with great speed.

screen memory: an alien abduction can be so terrifying and painful that an abductee's mind will block it out and replace it with a another one that's more tolerable. This new artificial memory is called a screen memory.

SETI: an acronym for the "Search for ExtraTerrestrial Intelligence." Established in 1960, SETI is a black project that's designed to cover up the fact that the governments of the world have long known that aliens are already here on Earth.

shadow government: another name for a black government.

sighting: an eyewitness sighting of a UFO or ET.

SIGINT: an acronym for "Signals Intelligence," a branch of Britain's intelligence community.

SIS: an acronym for Britain's "Secret Intelligence Service," also known as MI6.

skywatcher: another name for a UFO watcher.

sonar: an acronym for "sound navigation and ranging," sonar is a method for detecting, or communicating with, other craft, either underwater or in the air. There are two types: active (sending out sound pulses and listening for return echoes) and passive (simply listening for sounds). Numerous UFOs and USOs have been tracked by sonar.

space: begins sixty-two miles above the Earth's surface.

space sister/brother: one who believes he/she is linked with aliens through emotional feelings, genetics, or abduction.

SPADATS: an acronym for America's "Space Detection and Tracking System." Part of NORAD, SPADATS observes the airspace above the U.S. for enemy aircraft and missiles. As government documents show, on many occasions SPADATS has recorded genuine UFOs.

spiritualism: the belief that the living can communicate with the dead.

SPS: Security Police Squadron.

SSB: Soft Support Building.

STS: a NASA acronym for "Space Transportation System."

subculture: a culture inside or below another.

suborbit: less than one orbit.

sun dog: same as a mock sun.

supersonic: moving at speeds of between one and five times the speed of sound (741 mph).

target: something to aim at.

telemetry: the scientific process of collecting information, such as height, speed, pressure, or temperature. The measuring device that's used in telemetry is called a telemeter. After collecting the information, the telemeter transmits it by radio to a distant receiving station for analysis. Satellites often collect telemetry of UFOs, as they move around and through our skies.

telepath: one possessing extrasensory perception; one who communicates from one mind to another.

telepathic contactee: one who attains ESP after being abducted.

telepathy: communication between minds; most alien species seem to communicate between themselves and with humans telepathically.

teleportation: moving an object without physical contact; psychokinesis.

theodolite: an instrument used to measure vertical and horizontal lines and angles.

thought disk: a saucer-sized disk created by aliens by thought alone.

TIA: temperature inversion analysis.

TIF: an acronym for the "Theory of Intelligent Force." Some people explain crop circles, for example, by the TIF, believing that they're created by alien beings.

TLP: an acronym for "transient lunar phenomena," unknown lights that move on or above the Moon's surface; they've been seen by scientists for many centuries but have never been fully explained.

topography: mapmaking and/or surveying.

topology: scientific study of a physical place.

transonic: moving at approximately the speed of sound (741 mph at sea level).

TSA: an acronym for the U.S. "Transportation Security Administration."

UAO: unidentified aquatic object.

UAP: unidentified aerial phenomenon.

UAV: unmanned aerial vehicle; these remotely piloted aircraft function as spies in the air.

UFO: an acronym for "unidentified flying object," believed by most people (civilian, governmental, and military) to be extraterrestrial. Nearly all are unmarked, show no signs of propulsion, and can travel at speeds and perform maneuvers impossible for human-made aircraft.

ufocal: centered on or around UFOs.

ufoism: a belief in the reality of UFOs.

ufological: anything related to UFOs.

ufologist: one who scientifically studies UFOs.

ufology: the scientific study of UFOs.

ufonauts: same as aliens or ETs.

UFOria: a feeling of well-being or excitement that's brought on by the topics of UFOs and aliens.

UGM: an acronym for "unusual ground marking," and typically linked with UFO landings. Also another term for crop circles.

ultradimensional: beyond our material dimension.

undefined sensory experience: scientific term for the experiences abductees go through.

underwater unidentified object (UUO): an underwater UFO.

unexplained aerial object (UAO): same as UFO.

unidentified atmospheric phenomena: weather related phenomena that often appears to be, or can actually be, a UFO.

unidentified flying object: a UFO.
unknown: in aeronautics and ufology a UFO.
uranian: a being from the planet Uranus; or anything related to Uranus.
USAF: an acronym for America's "United States Air Force." Government documents show that the USAF, like many other government bodies, has been involved in UFO research for many decades.
USE: see undefined sensory experience.
USG: United States Government.
USO: an acronym for "underwater submerged object." Simply a UFO that has the ability to travel underwater.
UUO: an acronym for "underwater unidentified object." See USO.
velocity: the rate of motion.
venusian: a being from the planet Venus; or anything related to Venus.
vortex center: a portal between the physical and the spiritual worlds.
X-files: highly classified governmental documents pertaining to paranormal events.
zulu: used for the letter "z." Mainly British.

INDEX

THIS IS BOTH AN ORDINARY INDEX AND ALSO WHAT I CALL A "PAPER SEARCH ENGINE." I'VE INCLUDED KEYWORDS THAT ALLOW ONE TO SEARCH NOT ONLY FOR SPECIFIC TOPICS BUT ALSO FOR SPECIFIC WORDS.

AAC 225
AAF 225
abacus 174
abductee 83, 84, 88-90, 96, 225, 229, 242
abductees 82, 83, 87-91, 96, 144, 242, 252
abduction . 7, 81-85, 87-96, 101, 144, 197-199, 210-213, 225, 226, 242, 249, 250
Abduction In My Life (Maccabee) 197
abductions 9, 11, 84, 91, 93, 196, 198, 213, 230
above Top Secret 110, 121, 127, 199
Above Top Secret (Good) 199
acceleration 34
ACERN 206
acorns 26
acronyms 219, 225
ACUFOS 206
Adams, Shelby L. 344
Adamski, George 195
ADC 225, 226
Aditi 52
Advanced Research Projects Agency 226
advanced state of mind 230
AEC 225
Aerial Phenomena Research Organization 226
aerial ships 68
Aerobatics 225
Aerolite 225

Aerospace Defense Command 225
Aerospace Intelligence Division 237
aerospace technology 35
AFB 225
AFGWC 225
AFIS 225
AFOC 225
AFOSI 225
AFR 225
Africa 21, 62, 150, 173, 174
AFSS 225
Age of Flying Saucers 226
Agent Orange disaster 117
agriculture 53
Air Base Commanders 125
Air Defense Command 226
Air Force 11, 33, 40, 58, 73, 76, 81, 115, 120, 125, 136, 137, 145, 153, 154, 167, 184, 196, 225, 237, 247, 248, 253
Air Force Base 225
Air Force Global Weather Control 225
Air Force intelligence 33
Air Force Intelligence Manuals 125
Air Force Intelligence Service 225
Air Force Office of Special Investigation 225
Air Force Operations Center 225
Air Force Regulation 225
Air Force Security Services 225
Air National Guard 226
air raid sirens 71
Air Technical Intelligence Center 227
air temperature 189
air-battle 232
aircraft . . 11, 17, 22, 25, 27, 31-36, 44, 66, 67, 70, 71, 73-76, 83, 84, 88, 99, 103, 104, 106, 107, 121, 122, 139-142,

149, 151, 153, 156, 159, 164, 166, 167, 169, 221, 224, 226, 230-234, 243, 244, 246, 250, 252
aircraft carrier 25, 75
aircraft identification 76
airliner 26
airplane 25, 27, 34, 70, 76, 115, 122, 143, 224
airplane hangars 143
airplanes 22, 26, 27, 184
airport 77, 83
airports 25, 106, 162, 184, 191
airship, defined 226
airships 69, 80, 197
Alaskan Air Command 225
ALCIONE 208
Aldrin, Edwin "Buzz" 158
alien . 7, 9, 11, 18, 32, 35, 40, 45-50, 52, 54-56, 58, 59, 62-64, 66, 67, 69, 73, 79, 81-85, 87-94, 96, 100, 101, 106, 109, 111-114, 116, 119, 121-123, 126, 132, 150, 152, 156, 158, 163, 185, 187, 195-199, 202, 203, 207, 211-214, 225, 226, 232, 233, 236, 237, 239, 241, 242, 248, 249, 251
Alien (movie) 59, 79
alien abduction . . 7, 81-85, 87, 88, 90-94, 197, 211, 212, 242, 249
alien abductions 9, 11, 93, 196, 198
alien agenda 81, 82, 100, 226
alien animals 203
alien autopsies 116
alien bases 202
alien beings 40, 58, 93, 113, 114, 119, 150, 156, 251
alien civilization 45, 233
alien civilizations 111, 112, 132
alien implant 89, 236
Alien Liaison (Good) 199

alien races 237
alien spacecraft 9, 45, 106, 121, 122, 163, 197, 242
alien technology 88, 112, 126, 156, 212
aliens . . 1-4, 7, 10, 16-19, 27, 36-38, 44-46, 48-63, 66-68, 72, 78-84, 86-96, 100-103, 107, 109-120, 123, 125, 126, 137, 138, 144, 147, 150, 153, 155, 156, 164, 167, 176, 181, 183-187, 193, 194, 197-199, 212, 216, 220, 221, 223, 226, 228-231, 234-237, 240, 242, 248-252
alien-like creatures 62, 63, 66
alligator 50
Alpha Centauri 36
altitude 18, 42, 165, 166, 221, 223, 232
aluminum pan 30
AMB 226
ambassador 226
amber 29, 65, 223
America 2, 4, 24, 32, 34, 38, 51, 59, 61-63, 68-71, 73, 74, 83, 96, 98, 102, 104-106, 110, 113, 115, 116, 118, 120-122, 126, 136, 137, 144, 147, 153, 154, 159, 171, 174, 186, 187, 194, 196, 203, 210, 211, 225, 227, 228, 230-233, 236-240, 242-246, 248, 250, 253, 343
American military base 38, 187
American population 24
American president 110
American public 73, 115, 153
American soldiers 105
Americans 61, 63, 70, 83, 96, 113, 121, 126, 186
America's intelligence community 228, 231, 233
ammonia 55
amnesia 87, 226
amnesiac block 226
Anahita 67
Ancient Aliens (TV show) 79, 220
ancient myths 62, 142, 143, 173

Anderson, Loni 344
Andreasson Affair, the 94
Andreasson, Betty 93, 94
Andrews, Colin 200
Andromeda 36
ANG 226
angel hair 139, 140, 150, 189, 226
angels 62, 80, 93, 94, 198, 344
Angkor Wat 186
Anglo-Celtic society 2, 203
animal 11, 81, 97-101, 107, 173, 200, 225, 242, 245
animal mutes, books on 200
animal mutilations 11, 98, 99, 107, 200, 242
animal samples 81
animals . . 28, 39, 41, 90, 91, 97-101, 117, 119, 140, 141, 162, 166, 203, 230
annunciation 66
anomalies 226
anti-gravity machines 35
anti-matter system 35
antiaircraft guns 233
anti-gravity devices 36
ape-men 173
Aphrodite 52
Aphrodite's Trade (Seabrook) 204
Apollo 10 . . . . 147, 160
Apollo 11 . . . . 6, 147, 158, 160
Apollo 13 160
Apollo 14 . . . . 159, 160
Apollo 15 . . . . 155, 160
Apollo 16 . . . . 157, 160
Apollo 17 146
Apollo 8 . . . . 6, 157, 160
Apollo missions 158

apparitions 199
Applied Science Division 227
APRO 226
archaeology 201
Architeuthis dux 173
archives 28, 213
Archuleta Mesa 186
Arcturus 36
Area 51 . . . . 10, 121-123, 126, 187, 212, 213
Argentina 21, 206
Ariel School 150
Aristarchus 146, 147
Arizona 75, 76, 94, 97, 147, 186
Arkansas 75
Armstrong, Neil 158
Army Air Force 225
Arnold, Kenneth 72, 80, 187, 196
ARPA 226
art 61
Arthur, King 343
artillery guns 71
artwork 63, 68
ASD 227
ASEPI 207
Asherah 52
ashtrays 26
Asia 48, 66, 134, 135
Asians 48
asphalt roads 140
assistant 137, 237
Assistant Chief of Staff 237
assistants 47
Associação Pesquisa Ovni 209

Association Sciences de l'Etrange and Phenomenes Inexpliques 207
AST 227
Astarte 52
astrology 227
astronaut 66, 156-159, 227, 230
astronautics 45, 227
astronauts 6, 10, 147, 156, 158, 159, 168, 214, 241, 243
astronomer 51, 69, 137, 146, 148, 235
astronomers 21, 69, 80, 146, 147, 178, 241
astronomy 65, 111, 177, 247
astrophysics 227
ATIC . . 42, 47, 80, 84, 85, 105, 113, 122, 123, 129, 137, 141, 149, 150, 153, 163, 168, 181, 225, 227, 252
Atkins, Chet 344
Atlantic Ocean 227
Atlantic Standard Time 227
Atlantis 212, 214, 227, 238
Atlas missile 184
atomic 72, 80, 114, 184, 225
atomic bomb 72, 80, 114, 184
atomic bomber squadron 114
Atomic Energy Commission 225
Atthar 52
AUFORN 206
auras 227
Aurora 122
auroras 22
Australia . . . 21, 28, 62, 63, 83, 129, 130, 186, 199, 206, 213, 217, 247
Australian aborigines 63
Australian Centre for UFO Studies 206
Australian Close Encounters Resource Network 206
Australian UFO Research Network 206

authentic science 166, 175, 177
autokinesis 23
Aztecs 63
baby aliens 37
Babylon 5 (TV show) 79
Baca County, Colorado 102
back-engineering UFOs 122
back-engineered UFO technology 122
bacterium 233
Bailey, Francis 146
Balducci, Corrado 113
ball lightning 125, 166, 227
ball of light 240
balloons 11, 22, 23, 26, 27, 114, 125
balloons, shaped like UFOs 22
ballpoint pens 135
balls 26, 69, 71, 72, 75, 131, 187, 193, 201, 227, 234
balls of light 234
bananas 26
baptism of Christ 175
bar code 236
barium clouds 22
Barnes, Harry 73
Basel Broadsheet 151
Basel, Switzerland 151
basketball 25, 131
Bateson, Gregory 175
bathtub 25
Battlestar Galactica (TV show) 79
beam of light 40, 66, 82, 85, 95
bears 17, 90, 99
Beauregard, Pierre G. T. 344
Belgium 74, 206

believers . 9, 10, 51, 57, 76, 78, 131, 151, 152, 163, 165, 174, 178, 185, 187
bell 343, 344
bells 26, 69, 129
Bender, Albert K. 136
Bentwaters-Woodbridge Incident 140, 151
Berkner, Lloyd 118, 240
Berlin Seal 65
Berliner, Don 197
Berlitz, Charles 201
Bermuda Triangle 28, 201
Bermuda Triangle, books on 201
Bernstein, Leonard 344
Beta-UFO Indonesia 208
Beyond Top Secret (Good) 199
Bible 2, 63-65, 67, 82, 144, 196, 204, 343
bigfoot 187, 201, 202
Bilot, Maurice 181
binary-star system 92
binoculars 181, 182, 190
bird 122, 204
birds 22, 26, 32, 42, 90, 104, 140, 166, 203
Birmingham UFO Group 209
birth 28, 49
black budget 227, 228
black budgets 109, 110, 126
black Cadillacs 134
black disk 40
black eyes 46
black globes 151
black government 104, 107, 137, 228, 243, 246, 249
black governments 109, 110, 126
black hole 172
black holes 168, 172, 173, 175, 179

black programs 121, 126, 228
black project 99, 110, 228, 249
black projects 110
black suits 7, 133-136, 239
black ties 134
black UFOs 30
Blessed Lady of Ireland 2, 204
blimp 190
blimps 22
blip 228
blisters 106
blond hair 47
blood 2, 58, 87, 90, 91, 97, 100, 204, 344
blood drained 97
blood pressure 90
blood samples 87
Blot, Maurice A. 153
blue beam of light 95
blue eyes 47
BMW 228
body parts 98-102
Boeing 737 . . . . 17, 165
Boeing 747 25
Bogart, Humphrey 135
bogey 157, 228
bogies 203
Bok, Bart 178
Bolling, Edith 344
Bolton UFO Society 209
bomb . . 58, 71, 72, 80, 86, 114, 115, 123, 133, 183, 184, 228
Bomb Wing 228
bombs 58, 71, 183, 184
bone marrow 89
Bonilla, Jose A. Y. 178

Bonnybridge, Scotland 183, 191
boomerangs 26
Boone, Daniel 344
Boone, Pat 344
Bouchard, Pierre-François, Captain 248
Boulia, Australia 186
bowls 26, 131
boxes 26, 27
Bratton, Wiltshire, England 129
Brazil 63, 118, 155, 206
Brazilian UFO Research Network 206
Breckinridge, John C. 344
Bridge, the, on the Moon 148
Briggs, Katharine 203
bright lights 38, 75, 76, 83, 85, 91, 239
Britain 44, 78, 120, 154, 203, 233, 234, 238, 240, 249
Britain's intelligence community 229, 235, 247, 249
Britain's Ministry of Defense 231
Britain's Roswell 42
Britain's Security Service 239
Britannia Rules (Seabrook) 203
British government 40, 42, 144, 231
British intelligence community 239
British newspaper 129
British prime minister 110
British UFO Newsclipping Service 216
British UFO Research Association 210
British-American military base 38
Bronk, Detlev 117, 240
bronze 28, 29
Brooke, Edward W. 344
Brookings Institute 176
Brooks, Preston S. 344
Brown, Dan 204

Brown, Thomas Townsend 244
brownies 203
Bruno, Giordano 170
Buchanan, Patrick J. 344
Buffalo, Wyoming 102
BUFORA 210
Buford, Abraham 344
BUFOS 209
buildings 53, 122, 188
Bulgaria 207
bullet 44
bullets 26, 33, 112, 142
BURN 34, 76, 144, 206
burned at the stake 170
burning lamps 64
burns 38, 41, 88, 144
Bush, Vannevar 117, 133, 240
Butler, Andrew P. 344
B-2 Spirit 122
CAA 228
CAB 228
Cadillacs 134, 137
California 69-71, 97, 136, 184, 187, 189, 213, 225
call letters 103, 104
camera . 11, 17, 37, 39, 78, 134, 142, 159, 181-184, 188, 191, 233
cameras 37, 39, 78, 142, 181, 233
Campbell, Joseph 343
Campeche, Mexico 76
Canada . . . 21, 29, 98, 118, 129, 131, 202, 207, 211, 228, 234, 248
Canada's Roswell 29
Canadian 29, 123, 125, 131, 133, 199, 228, 247
Canadian Aviation Administration 228

Canadian Forces Station 228
Canadian government 125
Canary Islands 98
candlelight 170
cannon 71
cannons 59
canyon 68
Cape Canaveral, Florida 238, 243
capsule 157-159, 185
Captured! The Betty and Barney Hill UFO Experience (Friedman
. . . .)197
car 25
car accidents 187
car engines shutting off 141
car headlights 22, 236
carbon 147
cardboard 55
Carnton Plantation 2, 201, 345
Carnton Plantation Ghost Stories (Seabrook) 201
Carpenter, Scott 157
Carson, Martha 344
Carter, Jimmy 152
Carter, Theodrick "Tod" 344
Cascade Mountains 72
casebook studies 10, 38, 131, 149
casebook study . . . 13, 25, 26, 28, 29, 38, 56, 70-72, 74-77, 83, 92-94, 102, 105, 114, 118, 131, 135, 136, 149-151, 156-159, 184
Cash, Betty 105
Cash, Johnny 344
Cash/Landrum Incident 105, 106
Castle, the, on the Moon 147
catastrophe 227
Catholic Church 113, 169, 170

Catholicism 114
Catholics 52
cats 140, 166, 171
cattle 7, 9, 97-103, 105, 106, 117, 186, 200
cattle DNA 101
cattle mutilation 100-102, 200
cattle mutilations 9, 97, 98, 100-103, 105, 186
Caudill, Benjamin E. 343
CAUS 228
CBPDV 206
CCCS 211
CE1 . . . . 37, 44, 64, 65
CE2 . . . . 37, 44, 64
CE3 . . . . 38, 44, 64, 65, 67
CE4 . . . . 38, 44, 64, 66, 67
CE5 . . . . 38, 44
ceilings 56
cell phones 38, 170, 233
Celtic Mother-Goddess 2, 204
CENAP 207
Center for North American Crop Circle Studies 211
Center for Scientific Research 229
Center for Treatment and Research of Experienced Anomalous Traum 212
Center for UFO Studies 189, 210, 217, 230
Central Intelligence Agency 228
Central Security Control 230
Centre for Crop Circle Studies 211
Centre Nationale d'Études Spatiales 229
Centro Brasileiro de Pesquisa de Discos Voadores 206
Centro de Ufologia Brasileiro 206
Centro Italiano Studi Ufologici 208
Centro Ufologico Nazionale 208
cereologists 130, 132, 228

cereology 129, 132, 228
CERES 211
CERPA 207
certifiable UFOs 142
Cessna 182 83
CFS 228
CGS 228
chairs 26
chariot of fire 64
chariot of the Sun 62
Chasing UFOs (TV show) 220
Chassin, Lionel Max 177
Chatelain, Maurice 6, 158, 202
Chaves County, New Mexico 115
Cheairs, Nathaniel F. 344
cheese 55
cherubim 63
Chesnut, Mary 344
Chicago Department of Aviation 77
Chidlaw, Benjamin 97
child 49
children 51, 67, 68, 83, 91, 93, 150, 151
children's testimonies 150
Chile 207
chimpanzees 90
China 25, 63, 149, 151, 207
Chinese 55
Chinese lanterns 22, 125
Chop, Albert M. 73, 153
choppers 7, 97, 103, 200
Christ 2, 113, 175, 204
Christ Is All and In All (Seabrook) 204
Christian legend 83
Christian legends 67

Christmas Before Christianity (Seabrook) 204
Christmas tree tinsel 140
Chronological Catalog of Reported Lunar Events (NASA) . . 146
chthonic 246
chupacabra 213
church 68, 113, 169, 170, 185
Church of Jesus Christ of Latter-day Saints 113
Churchill, Winston 120
CIA 109, 145, 153, 228, 239, 245
CIA headquarters 229
CIC 229
CIFE 209
cigar-shaped UFO 26, 27, 69, 70, 80, 225, 229
CINC 229
cinnamon 55
CIO 207
cipher 229
circles . . . 7, 9, 11, 26, 129-133, 140, 200, 211, 212, 228, 251, 252
Circles Effect Research Society 211
Circles Phenomenon Research 211
Circles Research 211
circular patterns 33, 130
CISU 208
cities on the Moon 147
Citizens Against UFO Secrecy 228
Civil Aeronautics Board 228
Civil Rights Movement 74
civilian 28, 29, 72, 106, 137, 142, 167, 252
civilians 40, 48, 118, 120, 151
civilization 45, 154, 204, 233
clairaudient 229
clairvoyant 229
Clark, Jerome 195, 201

Clark, William 344
classifications of UFOs 37
classified 229
clay tablet 65
climb 18, 33, 36, 44, 94, 122, 157
cloaking 30, 77
clocks 88
Cloera, Ireland 68
close encounter 37, 38, 56, 242
Close Encounter of the Fifth Kind 38
Close Encounter of the First Kind 37
Close Encounter of the Fourth Kind 38
Close Encounter of the Second Kind 37
Close Encounter of the Third Kind 38
close encounters . . . 37, 56, 62-64, 68, 80, 92, 144, 145, 206, 212
cloud formation 238
clouds 17, 22, 26, 33, 64, 77, 146, 149, 236, 241
cluster UFO 229
CNACCS 211
CNES 229
CNRS 177, 229
Coast Guard Station 228
COBEPS 206
cockpit 228
coded writing 230
coelacanth 173, 174
coins 26
cold black eyes 48
cold metal table 85
Coleman, Loren 202
colonizing Earth, aliens 81
color . . 29, 30, 43, 59, 65, 102, 107, 121, 144, 149, 189, 221, 223, 242

colors . . . 30
colossal squid . . . 173
Columbus, Christopher . . . 28
Combs, Bertram T. . . . 344
comet . . . 233
comets . . . 146
COMINT . . . 229
Command Post . . . 230
Commander-in-Charge . . . 229
commercial pilots . . . 26, 29, 121, 142, 143
Committee for State Security . . . 237
communication . . . 54, 131, 229, 251
communication with the dead . . . 132
Communications Intelligence . . . 229
Communications Security . . . 229
Communique . . . 217
compass . . . 172
compasses . . . 141, 233
computer . . . 141, 216
computers . . . 38, 122, 170, 174
COMSEC . . . 229
CONAIPO . . . 208
Condon Committee . . . 200
cone-shaped beams of light . . . 26
Connecticut . . . 136, 174
conspiracy . . . 10, 11, 104, 111, 116, 118, 126, 200, 214, 246
conspiracy of silence . . . 104, 126, 246
constant fear . . . 91
constellation . . . 62, 93
constellations . . . 36
Contact (movie) . . . 79
Contact International UFO Research . . . 210
contactee . . . 212, 213, 219, 229, 251
contactees . . . 82, 96

contactology 230
Continuum 217
contrail 230
contrails 35
conventional aircraft 35, 230
conventional science 169, 177, 178
conventional scientists 65, 147, 169, 171, 176
Conwy UFO Group 210
Cooper, L. Gordon 156
Copernicus, Nicolaus 170
Copley Wood 199
cornea 23
Cornwall UFO Research Group 210
Corporacion para la Investigacion Ovni de Chile 207
Corso, Philip J. 156, 197
cosmic consciousness 230
Cosmic Conspiracies 210
cosmonauts 159, 230
Costilla County, Colorado 102
cotton candy 140
Counter-Intelligence Corps 229
country roads 24
court of law 152
covering up UFOs 112
coverup 2, 7, 109-111, 114, 117, 118, 120-122, 124-127, 137, 144, 154, 199
cow 99-101, 149
cows 39, 99, 140
coyotes 98, 99
CP 230
CPH 211
CR 211
craft . . . 9, 11, 17, 18, 22, 23, 25-36, 38, 39, 43-45, 47-49, 61, 63-77, 82-85, 87, 88, 92, 94-96, 99, 101-107, 110, 114,

115, 118, 119, 121, 122, 125, 139-143, 145, 147, 149-151, 153, 156-159, 163-167, 169, 184, 187, 189, 197, 221, 224, 226, 230-234, 242-244, 246, 249, 250, 252
crash site 114, 116
crashed UFOs 114, 122, 126
crater 146-148, 158
Crawford, Cindy 344
creature . 45, 46, 49, 50, 52, 57, 59, 61, 66, 68, 82, 119, 120, 158, 172, 233
creatures . . . 37, 38, 46, 53, 54, 60, 62, 63, 65, 66, 68, 69, 82, 83, 92-96, 98, 119, 150, 151, 154, 173, 177, 198, 202, 203
crescents 26, 27
crescent-shaped craft 149
Cretaceous Period 174
crimson 29
Crivelli, Carlo 66
Crockett, Davy 344
crop circle 129-131, 200, 211
crop circles . . 7, 9, 11, 129-133, 140, 200, 211, 212, 228, 251, 252
crop circles, books on 200
crosses 26, 86
Croste, M. 69
crucifix of Jesus 68
crucifixion 67
Cruise, Tom 344
cryptozoology 230
CSC 230
CUB 206
cubes 26
cucumbers 26
CUFOS 189, 206, 210, 230, 244
cult group 230

cultists 99, 107
CUN 208
cups 26
cusp 230
cuts 38, 88, 96, 98-100, 144
Cybele 52
Cygnus 146
cylinders 26
Cyrus, Billy R. 344
Cyrus, Miley 344
Czech Republic 207
Da Panicale, Masolino 82
Damascus 66
Däniken, Erich Von 203
Dark Skies (movie) 79
DATT 230
Davis, Jefferson 343, 344
daytime disk 231
Dayton, Texas 105
DCD 230
DCSOPS 230
DD 66, 182, 220, 231
DDO 231
DDs 25, 29, 43, 64, 74, 131
de Gelder, Aert 175
De Rostan 69
dead aliens 49, 116, 119, 156
dead cattle 97, 98
dead end science 177
debunk 78, 231
debunker 17, 231
debunkers 7, 51, 130, 152, 163-165, 167, 174, 176, 178
debunking UFOs 124
decaying meat 55

deceleration 34
deception 110, 111, 167, 198, 200
deer 90, 140
Defense Attache 230
Defense Intelligence Agency 231
Defense Intelligence Staff 231
Delgado, Pat 200
delta-winged 231
Demilitarized Zone 232
demonism 231
demonology 231
demons 137, 231
denial 104, 111, 117, 213, 246
Denmark 207
Denning, William Frederick 146
dentists 51
depression 89, 91
Deputy Chief of Staff for Operations and Plans 230
Deputy Director for Operations 231
desert 24, 64, 114-116
devil 119, 132
DIA 231
diamonds 26, 27
diamond-shaped UFO 105
diarrhea 91, 106
digital memory card 182
dinosaur 50, 203
Direction de la Surveillance du Territoire 232
Director General of the USAF 125
dirt 23
DIS 231
disc . 10, 11, 18, 22, 32, 38, 40, 42, 51, 58, 59, 66, 70, 79, 87,
89, 91-93, 97, 102, 106, 111, 115-117, 119, 123, 124,
134, 136, 139, 143-148, 150, 159, 166-168, 172, 173,

176, 178, 193, 194, 198, 200, 202, 206, 220, 231, 232, 243, 248
discarded space junk 22
dischi volanti 150
Discovery Channel 79
discovery of writing 70
discrediting military pilots 111
discs 115, 193, 231
disinformation 111, 112, 126, 231, 238, 245
disinformation programs 112
disk . 22, 27, 28, 40, 61, 68, 70, 104, 115, 140, 155, 157, 159, 175, 231, 251
disks 24, 67, 73, 80, 187, 231
disk-shaped objects 61, 159
disk-shaped UFO 157
District 9 (movie) 59
diurnal disk 231
dive 33, 36, 43, 44
divine beings 62, 143
divine messengers 63
dizziness 91
dizzy 140
DMZ 232
DNA 49, 50, 53, 81, 82, 89, 96, 100, 101, 107, 214, 226
DO 232
doctor 91, 144
doctors 91, 116, 144, 168, 215
documents 59, 78, 80, 110, 111, 117, 120, 121, 123-126, 144, 205, 228, 229, 231, 234, 240, 243, 245, 250, 253
dogfight 232
dogs 240
Dogu statues 66
dolphin 60
dolphins 166

dome 40
domes 30
Domestic Collections Division 230
double-star system 93
doughnuts 26
dove 28, 67
Dowding, Lord Hugh 139, 154
Dr. Who (TV show) 79
Draco 36
Dracos 186
Dreamland 212, 214
drone UFOs 31
drought 86
drug smugglers 76
DST 232, 233, 239
Dulce Base 186
Dulce, New Mexico 186
dumbbells 26, 129
dust 23, 86
Duty Officer 232
Duvall, Robert 344
dwarfs 203
E.T. The Extraterrestrial (movie) 59
early UFO sightings (list) 196
ears 17, 32, 90, 97, 99, 100
Earth 17, 21, 32-34, 37, 46, 49, 50, 52-54, 58, 59, 61, 63, 65, 66, 72-74, 77, 79, 81, 82, 86, 89, 93, 94, 100, 109, 112, 113, 123, 125, 129, 130, 132, 139, 143-145, 153, 155-157, 159, 162-164, 167, 169-171, 183, 186, 190, 200, 201, 203, 220, 226, 232, 233, 235, 237, 239, 241, 248, 249
Earth mysteries 129, 132
Earth, destruction of by our Sun 50
Earth: The Final Conflict (TV show) 79

Earth-Goddess 232
earthian 232
earthlight 232
earthling 232
earthquakes 86
Earth-Mother-Goddess 130
Earth's atmosphere 33, 34, 74, 165, 166, 239
Earth's gravity 44
Earth's satellite 241
Earth's weather 24
East 2 West UFO Society 209
East Indian myths 63
Easter Island 186
eating utensils 135
EBE 46, 232
echelon 32, 72, 149, 232
ECM 232
ecology, and the aliens 86
Editing, Debriefing, and Continuity Branch 237
Edward I, King 343
Edwards, Frank 203
eggs 26, 69
Egypt 248
Egyptian demotic 248
Egyptian hieroglyphs 61, 115, 248
Egyptians 62, 70
Einstein, Albert 36, 37, 44, 172, 176
Eisenhower, Dwight 156, 243
electric light bulb 145
electrical systems, UFOs shut off 112
electrically charged particles 73
electromagnetic effect 141
electromagnetic energy 232
electromagnetics 232

electronic countermeasures equipment 232
electronic devices 232
electronic equipment 37, 141, 162
electronic instruments 141
electronic intelligence 232
Elijah 64, 82
ELINT 232
elves 130, 132
EM 232
embarrassment factor 112
embryonic aliens 37
EME 141, 232
emergency flares 22
Encyclopedia Britannica 176
end-of-the-world scenarios, shown during abduction 86
energy-producing technologies 184
engine 25, 36, 39, 166, 216, 255
engineers 178
engines 35, 141, 226
England 38, 68, 69, 75, 98, 129, 146, 149, 154, 183, 185, 211, 216, 217, 343
enigma 29, 98, 129, 137, 195, 199, 200, 217
Enoch 64, 82
environmentalists 91
equipment, UFO watching 183
Erda 232
ESP 251
ET . . . 4, 11, 60, 117, 207-209, 212, 214, 215, 226, 230, 233, 236, 249
ETH 163, 233, 344
ETs 38, 79, 88, 113, 156, 186, 252
euphoria 193
Europe . . 2, 24, 69, 70, 98, 165, 173, 174, 203-205, 232, 246, 343

European Earth-Goddess 232
Europeans 70, 173
evidence 7, 11, 32, 37, 53, 73, 79, 90, 99, 106, 117, 139, 140, 144, 145, 147, 152, 156, 160, 162-164, 169, 171-175, 178, 189, 191, 196-198, 200, 201, 203, 212, 213, 235
exhaust 35, 36
exhaust openings 36
exhaust trails 35
Exodus 64
experiencer 51
experiencers 51, 53-55, 57
experimental military aircraft 22, 164, 166
exploding shells 233
explorers 28, 34, 173
expressways 24
extrasensory perception 233, 251
extraterrestrial . . . 6, 11, 29, 59, 61, 67, 72, 80, 81, 83, 88, 89, 101, 111, 114, 115, 123, 132, 139, 144, 153, 164, 176, 181, 187, 197, 199, 201-203, 232, 249, 252
extraterrestrial biological entity 46, 232
extraterrestrial creatures 83
extraterrestrial hypothesis 163, 233
extraterrestrial presence 144, 197
extraterrestrial spacecraft 11, 114
extraterrestrial technology 101
extraterrestrials 18, 49, 64, 66, 139, 225
eye 23, 25, 41, 96, 158, 162, 165, 166, 182, 196, 248
eyeball 23
eyeglasses 23
eyes 17, 23, 35, 46-50, 55-57, 60, 65, 85, 86, 93, 95, 97, 100, 105, 106, 119, 134, 136, 151, 173, 174, 187
eyewitness . 29, 51, 65, 84, 110, 117, 145, 151, 174, 186, 235, 249
eyewitness sightings 235

eyewitness testimonies 110
eyewitness testimony 29, 51, 84, 117, 151, 186
eyewitnesses . 22, 28, 29, 32, 35, 46, 48, 50, 55, 69, 75-78, 86, 101, 105, 107, 109, 116, 118, 120, 122, 130, 131, 136, 140, 145, 150-152, 164, 171, 188, 229, 232, 238
F-94 154
FAA 77, 103, 104, 233
Fact or Faked: Paranormal Files (TV show) 79
Fahrney, Delmer S. 155, 163
fairies 130, 132, 203
fairy rings 130
fake documents 111
fake UFO 235
falling leaf 33
false science 7, 163, 170-172, 174, 175, 205
false scientists 171, 174, 178
fame 51, 168
FAO 206
Far East 21
farm 39, 41, 99, 129
farmers 97
Fast Walker 74
Fate magazine 217
Father of Modern Rocketry 154
Fawcett, Lawrence 199
FBI 115, 123, 124, 145, 193, 199, 232, 233, 239
fear of doctors 91
fear of the unknown 176
Federal Aviation Administration 233
Federal Bureau of Investigation 233
female 52, 53, 60, 62, 63
female leader 53
Feokistov, Konstantin Petrovich 159
Ferris wheel 150

feu . . . 71
fiber optics . . . 122
fields . . . 7, 25, 26, 66, 76, 97, 129-131, 140, 174
fighter jets . . . 76, 105, 157
files . . . 11, 25, 29, 38, 79, 117, 124, 149, 194, 199, 213, 220, 244, 253
film 40, 49, 109, 134, 136, 142, 147, 151, 158, 182, 184, 188, 189, 344
film companies . . . 182
films . . . 46, 79, 92, 125, 221, 224, 344
finger . . . 85, 87
fingers . . . 46, 49, 85
Finland . . . 207
fire . . . 22, 28, 35, 39, 50, 51, 53, 61, 64, 65, 67, 86, 95, 118, 119, 125, 143, 151, 166, 174, 187, 193, 198, 201-203, 233
fire circles . . . 62
fire department . . . 119
Fire in the Sky (Walton) . . . 95
fireball . . . 151, 233
firefighters . . . 119, 120
fish . . . 90, 174
Flagstaff, Arizona . . . 147
flak . . . 233
flame . . . 29, 105
flames . . . 35
flame-colored . . . 29
flap . . . 72, 185, 233
flashing lights . . . 30
flashlights . . . 38, 41
flattened grass . . . 189
flight . 17, 69, 70, 87, 157, 159, 167, 184, 185, 191, 201, 226, 232, 234, 242, 243
Flight 19 . . . 201

float 18, 44, 70, 85
floating 56, 68, 85, 245
floods 86
flying carts 63
flying craft 232
flying disk 115, 140
flying machines 63, 99, 170
flying roll 65
flying saucer 115, 124, 126, 136, 196, 200, 217, 233, 238
Flying Saucer Digest 217
Flying Saucer Review 217
Flying Saucers and Science (Friedman) 197
flying wing 233
flying wing design 233
FOCONI 207
fog 39, 71
FOIA 122-124, 126, 144, 234
FOIA request 123
foo fighters 71, 234, 245
food 55
footage 49, 213
football fields 25, 76
footballs 26
Foote, Shelby 343
Forbes, Christopher 344
force fields 26
Ford, Gerald 153
forest 4, 17, 38-40, 118, 119, 185
forests 24, 94, 181
formation 61, 72, 75, 232, 234, 238
formation of aircraft 232
Forrest, Nathan B. 343, 344
Forrestal, James 117, 240
Fort Riley, Kansas 156

fossil record 54, 247
Foundation for Cosmonoetic Investigations 207
four fingers 46, 49, 85
Fowler, Raymond 94, 194, 198
France 36, 110, 146, 177, 207, 229, 239
France's intelligence community 232
fraud 235
Freedom of Information Act 124, 234
French Air Forces 177
French astronomer 69
French government 235
French Secret Service 154
French soldiers 68
fresco 67
freshly plowed fields 97
Freud, Sigmund 91
Freya 52
friction 34, 165, 239
Friedman, Stanton T. 15, 153, 197, 199, 205, 342
Fuhr, Edwin 131
Fuller, John 93
Fund for UFO Research 210
Fundación Anomalía 209
Fundación Argentina De Ovnilogía 206
Fundacja Nautilus 209
fuselage 234
fuzzy 26
F-117 Nighthawk 122, 139
F-22 Raptor 122
F-35 Lightning II 122
F-94 fighter jet 154
Gaelic 54
Gagarin, Yuri 146
Gaia 130

galaxies 190, 234
galaxy 45, 75, 154, 167, 170, 234
Galileo 65, 170
Gansu Province 151
GAO 116, 117, 197, 234
GAO Report on Roswell 116, 117, 197
garbage cans 26
gases 22, 99, 147, 149
Gayheart, Rebecca 344
GCHQ 235
GCI 235
geese 98
Gemini 4 . . . . 157, 160
Gemini 7 157
Gemini missions 158
Gemini space capsule 185
gene 2, 79, 343
General Accounting Office 234
General Electric 153
genes 47
Genesis 63, 64
genetic 50, 60, 82, 100, 186, 215, 226, 250
genetic experiments 186
genetic material 50, 82, 100
genetics 250
geocentrism 169
GEPAN 235
Germany 62, 69, 156, 207
Ghirlandaio, Domenico 68
ghost 2, 66, 68, 85, 195, 201, 235, 343, 345
ghost rockets 70, 195, 235
ghost stories 2, 201, 343, 345
ghosts 11
giant insects 48

Giant Rock, California 187
giant squid 172, 173
Gibbs-Smith, Charles Harvard 154
Gilson, Case 225
Gist, States R. 344
glass 30
glassy strands 150
gliders 230
globe-shaped UFO 149
Glossary 224
glowing disk 68, 104
glowing gases 149
GMT 235
god 52, 62, 63, 131, 158, 177, 204, 205, 212, 214, 237
goddess . . . 2, 52, 53, 60, 62, 63, 114, 130-132, 200, 203-205, 214, 232
Goddess-worship, among Mormons 114
Goddess-worship, and crop circles 130, 132
goddesses 63, 205
gods 63, 67, 201-203, 212, 215
Goldwater, Barry 120, 121, 153
golf ball 25
Gollnow, H. 178
Good, Timothy 15, 38, 195, 199, 342
Gorbachev, Mikhail 155, 159
Gordon, George W. 344
gorilla 172, 173, 179
gorillas 172, 173, 179
Gospels 66, 204
Gould, Rupert T. 202
Government Accountability Office 234
government agencies 111, 234
government agency 134, 227, 235, 243
government agents 138

Government Communications Headquarters 235
government documents 110, 117, 123, 124, 126, 144, 228, 231, 243, 245, 250, 253
government employees 116, 122, 143, 218
government official 44, 116
government officials 73, 111, 123, 152, 162, 227
governmental documents 144, 253
Graves, Robert 204, 343
gravitational fields 26
gravitational forces 48
gravity 34-36, 44, 48, 164, 172, 214
gravity amplifier 35
Gray . 29, 46, 77, 85, 87, 93, 95, 117, 159, 165, 200, 223, 240
Gray, Gordon 117, 240
Grays 46-48, 53-55, 57
Great Blackout 141
Great Britain 238
Great Britain, UFO sightings increasing in 78
great Celtic Mother-Goddess 2, 204
Great Pyramid 186
Great Red Spot 164
Great Siberian Explosion 203
Greek 248
Greek mythology 228, 246
Greek myths 62
Greeks 70
green lights 40
green liquid 56
Greenwich Mean Time 235
Greenwood, Barry J. 199
GREPI 209
Groom Lake, Nevada 121, 187
Ground Control Intercept 235
Ground Saucer Watch 235

ground-based radar 142, 145
Groupe d'Études Phénomènes Aerospatiaux Non Identifiés . 235
Gruythuisen 69
GSW 235
Guaraldi, Vince 344
Guérin, Pierre 177
guided-missile program 155
Gulf Breeze UFO Conference 218
Gulf Breeze, Florida 74, 130, 186, 191
Gulf of Maine 29
Gulf War Syndrome 117
guns 58, 59, 71, 97, 112, 233
Gutierrez, Pedro 28
gynecologists 93
g's 34
H-bomb 133
hair loss 106
hair samples 87
half-circles 26
Halley, Edmund 69
hallucination 164, 167, 169, 179, 182
hallucinations 51, 139, 163, 167, 168, 178
Halstead, Frank 178
Halt, Charles 40-42
Hangar 1 (TV show) 79
hard evidence 162, 213, 235
hard sightings 235
Harding, William G. 344
hats 26, 134
Haunted Highway (TV show) 79
Hawaii 62
Hawking, Stephen 156
hazardous weather 71
headaches 91, 106

health problems 91, 92, 96, 106, 107
healthy skepticism 175
heaven 63, 64, 67, 202
Heavenly Mother 52
heavens 22, 67, 173, 190, 233
heavily populated areas 88, 149
Hebrews 64
Heflin, Rex 136
helicopter 99, 102, 105, 222, 224, 242, 243
helicopters 10, 22, 39, 70, 98, 101-107, 186, 200, 236
heliocentrism 170
helmeted space beings 62
Hera 63
Hermes 63
Herodotus 147
Hess, Seymour 178
Heuvelmans, Bernard 203
hexagons 26
hieroglyphs 248
high strangeness 98, 235
Hill, Barney 92, 93
Hill, Betty 92
Hill-Norton, Lord 155
Hillenkoetter, Roscoe 117, 153, 240
Hiroshima, Japan 72
history 246
History Channel 79
History's Mysteries (TV show) 79
Hitching, Francis 200, 203
Hoagland, Richard C. 147, 201
hoax 164, 167, 170, 213, 222, 224, 235
hoaxers, UFO 22
hoaxes 51, 130, 131, 163, 167, 168, 178, 236
hoaxing 111, 126, 245

hoaxing UFO sightings 245
hoaxing, UFOs 167, 236
hobgoblins 203
holes 39, 88, 97, 144, 168, 172, 173, 175, 179
homing mechanism 90
Homo erectus 47, 247
Homo sapiens sapiens 47, 53
Hong Kong UFO Club 207
Hood, John B. 344
Hoover, J. Edgar 115
Hopkins, Bud 198, 199
horse 61, 195, 215
horses 63, 64, 140
Horseshoe Lagoon, Tully, Queensland, Australia 130
horseshoes 26
hospitals 91
hot gases 22
hot spots 183-186, 188, 235, 236
hot steam 35
house lights 22
House of Representatives 120
houselights 22
Houston Control Center 243
hover . 32, 33, 35, 36, 44, 71, 83, 99, 121, 122, 142, 149, 229
hovering . . . 21, 39, 61, 64, 70, 72, 74, 75, 77, 84, 85, 95, 99, 103-105, 131, 150, 155, 175, 181, 248
hovering spacecraft 64, 85
Hull UFO Society 210
human . . 17, 23, 25, 33, 34, 39, 44, 46-50, 54, 58, 60, 66, 69, 71, 81, 82, 88, 89, 91, 95-97, 100, 101, 104, 107, 113, 118, 130, 132, 139, 141, 149, 163, 167, 171, 184, 185, 190, 198, 199, 202, 215, 221, 222, 224-227, 230, 232, 236, 239, 248, 252
human being 236

human body 23
human DNA 81, 100, 101, 107, 226
human eye 25
human intelligence 236
human language 54
human samples 81
human society 81, 113
human-made aircraft 34, 230
humanoid 47, 49, 52, 236
humanoids 62, 198
humans . . . 18, 22, 27, 28, 32, 34-38, 46-50, 52-54, 56, 59, 60, 62, 63, 70, 77, 82, 84, 88, 89, 95, 96, 99, 101, 117, 130, 137, 138, 140, 146-148, 154, 155, 165-167, 171, 177, 186, 193, 230, 232, 236, 240, 246, 251
human-made aircraft . . . 17, 33, 34, 44, 71, 104, 139, 141, 149, 167, 221, 224, 230, 252
HUMINT 232, 236
Hunsaker, Jerome 117, 240
hurricanes 86
hybrid 49, 50, 82, 96, 101, 226, 236
hybrida 50
Hybrids 49, 50
hydrogen 147
Hynek, J. Allen 51, 137, 171, 172, 194-196, 235
hyperdimensional physics 148
hypnosis 88, 92, 102, 236
hypnotic regression 57, 82, 94, 95, 236
hypnotic techniques 135
hypnotist 88
hysteria 51, 113, 164, 167, 168, 178, 179
H-bomb 123
ICAR 210
ice 30
ice-cream cones 27

Idaho 97
identified flying object 22, 236
identifying aerial craft 125
IF 212
IFO 236
IFSB 136
IFUFOCS 212
Illinois Mutual UFO Network 210
illness 226
illnesses 38, 227
imagination 52, 57, 166, 175
immortal beings 63
implant 89, 90, 225, 236
implants 88, 89, 96, 144, 212, 213
Implications of a Discovery of Extraterrestrial Life 176
impressions 39, 130, 140, 182, 189
imprints 37
In Search Of . . . (TV show) 220
incisions 96, 99, 100
Independence Day (movie) 59, 79
Independent Alien Network 207
India 28
Indian Ocean 238
Indonesia 208
indoor plumbing 170
infants 51
INFRA 209
infrared light, UFOs operating in 35
insect 82
Insectile 48
Insectiles 48
insects 46, 48, 85
Institute for Atmospheric Physics 155, 171
Institute for UFO Contactee Studies 212

intake ports . . . 36
integrated circuit . . . 122
intelligence files . . . 234
interdimensional . . . 236
Intergalactic Federation . . . 237
International Center For Abduction Research . . . 210
International Flying Saucer Bureau . . . 136
International MUFON Symposium . . . 218
International Space Station . . . 139
international TV . . . 40, 49, 77
International UFO Congress . . . 210, 218
International UFO Reporter . . . 217
interplanetarians . . . 237
interplanetary . . . 227
interstellar matter . . . 234
INYSA . . . 237
INZ . . . 237
INZA . . . 237
ion clouds . . . 22
Iowa . . . 97
Ipswich, Suffolk, England . . . 38
IRAAP . . . 210
Iraq . . . 62
Ireland . . . 2, 62, 68, 186, 204, 208
Irish . . . 54
ironing boards . . . 27
Isis . . . 52, 62
Israel . . . 62
It Came From Outer Space (movie) . . . 59
Italy . . . 62, 150, 208
ivory . . . 29
I-beams . . . 115
J. Allen Hynek Center for UFO Studies . . . 230
Jackson, Andrew . . . 344

Jackson, Henry R. 344
Jackson, Thomas "Stonewall" 344
JACL 237
James, Frank 344
James, Jesse 344
JANAP 143, 237
Japan 62, 72, 129
Japanese 71
Japanese fighter planes 71
JCS 237
Jefferson, Thomas 344
Jehovah 237
jelly 46
Jent, Elias, Sr. 343
Jesus 2, 28, 63, 66-68, 82, 113, 175, 204, 230, 343, 345
Jesus and the Gospel of Q (Seabrook) 204
Jesus and the Law of Attraction (Seabrook) 204
jet . . . 32, 34, 43, 73, 104, 141, 142, 154, 157, 165, 168, 222, 224
Jet Propulsion Laboratory 237
jetliner 25, 142, 165
jets . . 22, 35, 58, 70, 76, 77, 87, 105, 112, 122, 139, 141, 142, 157, 165, 169, 170, 174, 183, 184, 230, 233
Jewish mythology 64
Jews 62, 70
John E. Mack Institute 212
John the Baptist 175
John, Elton 344
Joint Army-Navy-Air Force Publication 237
Joint Chiefs of Staff 237
jokes 167
Jomon, the 66
Journal of UFO Studies 217
JPL 237

Judd, Ashley 344
Judd, Naomi 344
Judd, Wynonna 344
Judge Advocate General 237
July 2, 1947 . . . . 114, 126
Jung, Carl 154
Juno 52
Jupiter 164
Jupiterian 237
Kaku, Michio 201
kaleidoscope 30
Kali 52
Kansas 97, 156
Kayapo 63
Kecksburg UFO Incident 118
Kecksburg, Pennsylvania 118
Keel, John 202
Kelle 2, 52, 204
Kennedy Space Center 184, 238
Kennedy, John F. 74, 81
Kennedy, Robert 74
Kentucky MUFON 210
Kevlar 122
Keyhoe, Donald 164, 196, 200, 244
KGB 111, 145, 213, 237, 238
Khrunov, E. V. 159
kidnap 96
kidnaped 225
kidnaping 50, 225
kilometer 238
King Island, Tasmania 83
King James Bible 65
King Thutmose III 62
King, Martin Luther 74

KISR 238
kites 22, 27
Kitty Hawk, North Carolina 69, 146, 166
KM 238
knapsack 182
knight 197
Komarov, Vladmir M. 159
Komitet Gosudarstvennoi Bezopastnosti 237
Korean War 71, 234
KSC 184, 238
Kuwait Institute for Scientific Research 238
La Esfera Azul 208
laboratories 174, 243
ladders 129
Lagarde, Fernand 110
lakes 26, 181
Landrum, Colby 105
Landrum, Vickie 105
Langelaan, George 154
Langenburg, Saskatchewan, Canada 131
Langley, Virginia 229
language 54, 55, 64, 65
lanterns 22, 27, 125
LAPIS 210
large metropolitan cities 149
Las Vegas, Nevada 76
laser 22, 38, 40, 62, 66, 98, 99, 106, 238
laser beam 98, 99
laser beams 106
laser displays 22
laser pens 38
laser-guided missiles 238
lasers 99, 100, 122
Latin 50

Latvia . . . 208
Launch Control Facility . . . 238
law enforcement . . . 234
law of physics . . . 44, 175
lawn mower . . . 32
laws of Nature . . . 173
laws of physics . . . 28, 33, 35, 37, 60, 71, 154, 168, 169
laws of science . . . 175
Lazar, Bob . . . 45, 213
LCF . . . 238
LDS . . . 113
Le Comité Belge pour l'Étude des Phénomènes Spatiaux . . . 206
leaves . . . 109, 140, 144, 183
Lee, Fitzhugh . . . 344
Lee, Robert E. . . . 344
Lee, Stephen D. . . . 344
Lee, William H. F. . . . 344
Lemuria . . . 238
lens . . . 238
lenticular cloud . . . 238
Leonard, George H. . . . 202
Les Repas Ufologiques . . . 207
Lewis, Meriwether . . . 344
Lewis, Steve . . . 33
ley line . . . 238
LGM . . . 238
license plates . . . 134
lie-detector . . . 58, 95
lie-detector tests . . . 95
lifeforms . . . 53, 113, 114, 190
lifelong observation . . . 89
light . . 17, 22, 25-28, 30, 31, 33, 35-37, 39-41, 43, 44, 46, 47, 56, 62-64, 66, 67, 69, 71, 75-78, 80, 82, 83, 85, 87, 93, 95, 101, 102, 105, 117, 118, 131, 132, 139, 141, 143,

145-147, 172, 173, 175, 184, 186, 187, 199, 214, 223, 227, 234, 240
light bulb 46
light bulbs 139, 141
light spectrum, UFOs and 35
lightbeam 87
lighthouses 22
lightning 22, 65, 67, 122, 125, 165, 166, 227
lights 17, 22, 26, 28-30, 38-43, 58, 63, 66, 69, 71, 75, 76, 80, 83, 85, 91-93, 98, 103, 107, 122, 135, 141, 142, 145-147, 158, 167, 170, 174, 183, 187, 194, 201, 221, 223, 224, 233, 239, 241, 244, 251
limousines 134
Lintiao Air Base 151
lips 97, 100
little green men 238
lizard 50
Loch Ness 202
Loch Ness Monster 202
Lockheed U-2 122
London UFO Studies 210
London, England 38, 146, 149, 154
Longstreet, James 344
Lord Kelvin 170
Los Alamos Atomic Energy Commission Project 184
Los Alamos, New Mexico 184
Los Angeles, California 71
loss of memory 226
love 2, 47, 86
Loveless, Patty 344
Lovell, James 6, 157
Lowell Observatory 147
LUFOS 210
Luke Air Force Base 76

Lunar 238
Lunar Orbiter IV 160
Lunar Orbiter V 160
Lunarian 238
Lyra 36
MAARS 210
MacArthur, Douglas 59, 81, 155
Maccabee, Bruce 15, 123, 124, 197, 199, 342
Machu Picchu 186
Mack, John 199
Mackal, Roy P. 202
Madonna 68
Madrid, Spain 149
magazines 78, 111, 182, 215, 216, 220, 231
magic 98, 186, 200, 215
magical energies 186
magnetic 25, 159, 171, 214, 238
mainstream science 172, 174, 176
mainstream scientist 152
mainstream scientists . . . 37, 61, 164, 166, 168, 174, 175, 178
Majestic-12 240
Majestic-12 Conspiracy 118
MAJIC 199
Malaysia 52, 208
Malaysian UFO Network 208
Malta 69, 186
man . . 10, 35, 48, 61, 62, 64, 67, 69, 116, 136, 146, 187, 201, 202, 212, 214, 226
Manigault, Arthur M. 344
Manigault, Joseph 344
manikin 68
manned flight 69
map of Nevada 123
mapmaking 251

marble 25
Marchetti, Victor 109
Marden, Kathleen 205
Mare Crisium crater 148
Marfa Lights 201
Mari/Meri 52
Mark 66, 87, 88, 144, 203
maroon 29
Marrs, Jim 197
Mars 22, 59, 193, 201, 212, 214, 215, 237, 238
Marseilles, France 146
Martian 238
Marvin, Lee 344
Mary 52, 67, 82, 83, 344
mass hysteria 113, 168, 179
mass hysteria factor 113
mass sighting 148, 238
massive coverup 126
Masters of Silence 111
material plane 137
mathematical formulas 129
Matthew 66, 67
mattresses 27
Maunder, Edward W. 178
Maury, Abram P. 344
MC 2, 120, 153, 155, 157, 158, 171, 344
McCormack, John W. 120
McDivitt, James 157
McDonald, James 155, 171
McDonnell Douglas 153
McGavock, Caroline E. (Winder) 344
McGavock, David H. 344
McGavock, Emily 344
McGavock, Francis 344

McGavock, James R. 344
McGavock, John W. 344
McGavock, Lysander 344
McGavock, Randal W. 344
McGraw, Tim 344
meat 55, 60
medical equipment 48
medical evaluation detector 90
medical operations 49, 98, 101
medical tests 57, 60, 82, 91, 93, 94
Medieval scientists 178
memory cards, camera 134
men . . . 9, 28, 39, 40, 48, 59, 66, 70, 77, 79, 83, 95, 97, 117, 118, 120, 133, 135-137, 139, 172, 173, 178, 200, 203, 238-240
men in black 9, 59, 79, 133, 135-137, 139, 200, 239
Men in Black (movie) 59, 79
Menippus 62
Menzel, Donald 117, 240
Mercurian 239
Mercury 239
Mercury 7 . . . . 156, 157
Mercury 8 . . . . 6, 157, 160
Meriwether, Elizabeth A. 344
Meriwether, Minor 344
Merlin C26A bimotor airplane 76
Mesonychoteuthis hamiltoni 173
Messier, Charles 69
metallic objects 142
meteor 118, 166, 174, 190, 233, 239
meteorite 17, 225, 239
meteorites 141, 173, 174
meteoroid 239
meteors 11, 22, 125, 139, 172-174, 179

Mexican Air Force 76
Mexican Air Force pilots 76
Mexico . 33, 49, 63, 76, 77, 102, 103, 114, 126, 150, 179, 184-186, 191, 208, 239
Mexico City, Mexico 150
Mexico's Department of Defense 77
MI5 . . . . 232, 233, 239
MI6 . . . . 145, 239, 249
MIB 7, 133-138, 239
MIB, books on 200
Michelangelo 68
Michigan 210, 248
Michigan MUFON 210
microwaves 170, 174
Middle Ages 67, 70, 80
Middle East 246
Midwestern states 97
military . . . 2, 11, 22, 23, 25, 28, 29, 32-34, 38-40, 42, 45, 48, 49, 58, 59, 70, 71, 73, 74, 76, 77, 79, 88, 101, 103, 105, 106, 110-113, 115, 116, 118-122, 126, 137, 140-143, 145, 148, 151-153, 162, 164, 166-169, 182-187, 191, 197, 199, 212, 216, 218, 228, 234, 236, 237, 244, 252, 343
military bases 25, 106, 183, 191, 212, 236
Military Channel 79
military documents 59
military installation 184
military jets 77, 142
military medics 39
military personnel . . . 11, 23, 38, 103, 118-120, 168, 218, 237
military weapons 110
milky way galaxy 167, 170
Millennium 239
Min Min Light 186

Ministry of Defense 240
Minneapolis, Minnesota 157
Minnesota 97, 157, 185
mirage 239
mirages 22, 125, 139, 239
missile 44, 155, 163, 184, 185
missiles 59, 112, 114, 142, 183, 238, 243, 250
missing time 87, 96, 141, 198, 240, 242
missing time experience 242
Mission Control 6, 158
Mission To Mars (movie) 59
Mitchell, Edgar D. 159
Mithras 67
MJ-12 . . . . 101, 117, 118, 123, 152, 199, 200, 240
mock sun 250
mock suns 240
MOD 152, 240
modern age 69, 80, 226
modern Age of UFOs 69, 80, 226
modern humans 53, 54
molten metal 40
monastery 67, 68
Monster Quest (TV show) 79
monsters 173, 179, 202
Montague, Robert 118, 240
Montana 97, 248
monuments 193, 201
Moon . . . 2, 6, 22, 30, 42, 62, 69, 80, 145-148, 150, 155, 157, 158, 166, 202, 213-215, 221, 224, 238, 241, 242
Moon crater 146, 158
Moon, artificiality of our 241
moonlights 148
Morgan, John H. 344
Mormonism 114

Mormons 52, 63, 113, 114
Mormons, and Goddess-worship 114
Morton, John W. 344
Mosby, John S. 344
Moscow, Russia 149
Moses 64
mother ship 242
mother ships 25, 242
Mother-Goddess 2, 52, 53, 114, 130, 204
Mothman 203
motor-powered airplane 70
Mount Rainier, Washington 72, 187, 191
mountain 66, 69, 80, 99, 122, 219
mountain lions 99
mountains 24, 72, 92, 122, 175, 187
movie camera 39, 184
movie star 47
movies, UFO 79
MQ-9 Reaper 31
Mt. Rainier, Washington 72
MTE 88, 242
mud holes 97
MUFON 189, 194, 210-212, 217, 218, 242
MUFON UFO Journal 217
Mulder, Fox 152
multiple abductee 242
multiple sighting 145, 168, 242
Museum of the Unexplained 210
Musgrave, Story 158
mushrooms 27
mutes 7, 97
mutilations 9, 11, 97-103, 105, 107, 186, 200, 242
mutilators 98, 107
mutologists 99, 242

mutology 107, 242
Mutual UFO Network 189, 210
mysterious lights, books on 201
mysterious, the 176
mystery . 7, 9, 11, 27, 34, 79, 97, 98, 100-107, 131, 133, 137, 177, 186, 195, 196, 198, 200-202, 204, 214, 215, 242, 243
mystery helicopter 105, 242, 243
mystery helicopters 102-107, 186, 200
mystery helicopters, books on 200
Mystery Quest (TV show) 79
mystical vibrations 186
myth 2, 62, 63, 202, 204, 205, 343, 345
mythology 52, 64, 177, 203, 228, 246
myths 62, 63, 142, 143, 173, 204
NAICCR 211
Napier, John 202
NARCAP 210
NASA . 6, 79, 146-148, 157-159, 161, 184, 202, 212, 213, 228, 237, 241, 243, 250
NASA photo technicians 148, 241
NASA transmissions 213, 241
NASA's Unexplained Files (TV show) 79
natalis 49
National Aeronautics and Space Administration 153, 243
National Aviation Reporting Center on Anomalous Phenomena 210
National Combat Operations Center 243
National Enquirer 182
National Geographic Channel 79
National Investigations Committee on Aerial Phenomena . . 244
National Investigations Committee on UFOs 218
National Military Command Center 244
National Photographic Interpretation Center 244

National Reconnaissance Office 245
National Science Foundation 245
national security 229
National Security Agency 245
National UFO Conference 218
National UFO Reporting Center 189, 210
Native American 63, 104, 187
Native American myths 63
Native Americans 61, 63, 186
Nature . . . 2, 22, 53, 104, 109, 148, 149, 169, 173, 176, 202, 204, 227, 235, 238
nausea 38, 91, 106
nautical mile 244
Naval Research Laboratory 244
Naval Space Surveillance System 243
navel 93
navigation lights 30, 43
navigation systems 31
NAVSPASUR 243
Navy 59, 123, 145, 155, 237, 244
Nazca 61
Nazca, Peru 61
NCOC 243
NDE 244
Near East 21, 246
near-death experience 244
Nebraska 97
nebulae 234
Need to Know (Good) 199
needles 91
Neonate 49
Neonates 49
NEPA 244
Neptune 244

Neptunian 244
Netherlands 208
NEUFOR 211
Nevada Triangle 121
New England 75, 98, 211
New England UFO Research Organization 211
New Hampshire 92
New Mexico 33
New Testament 2, 66, 80, 204
New York 56, 156, 212
New Zealand 208
newborn human baby 46
Newgrange, Ireland 186
news clipping services 215
news crews 182
news department 190
newspapers 72, 111, 115, 182, 191, 215, 231, 245
NICAP 164, 244
NICUFO 218
Night Skies (movie) 59, 79
night vision equipment 122
nightmare 57
Nightmare Hall 186
nighttime disk 244
nighttime lights 25
ninety-degree angle 121
ninety-degree turns 34, 122, 166
NL 66, 208, 244
NLs 25, 29, 43, 64, 69, 74, 131
NM 244
NMCC 244
nocturnal disk 244
nocturnal light 66
nocturnal lights 25

nonbelievers 51, 152, 165, 174
nonbelieving scientists 177, 178
NORAD 244
Nordic 47, 48, 63
Nordic aliens 48
Nordic myths 63
Nordics 47, 48
North America 62, 211, 248
North American Aerospace Defense Command 244
North American Institute for Crop Circle Research 211
North Pole 21, 186
Northern Lights 63
Northern UFO News 217
Northrop Grumman B-2 233
Northrop XB-35 233
Norway 21, 47, 208
Norwegians 47
Norwich, England 183
Nostradamus 68
notebook 182, 189-191, 219
Nova Scotia 29
NPIC 244
NRL 244
NRO 245
NSA 245
NSF 245
Nuclear Energy for the Propulsion of Aircraft 244
nuclear missiles 183
nuclear power 184, 191
nuclear power plants 184, 191
nuclear war 54, 86, 226
nuclear weapons 114
Nugent, Ted 344
numbness 91

nuns 51
nurses 116
OAM 245
OBE 245
Oberth, Herman 45, 154
Objects Seen Floating 245
oblong heads 48
occult powers 137
occult, the 245
ocean . . . 28, 29, 34, 62, 83, 84, 86, 145, 166, 184, 185, 227, 238
odd lights 98
office of disinformation 238
Office of Naval Intelligence 245
Office of Scientific Investigations 245
Office of Special Investigations 245
Office of Strategic Services 245
official agencies 231
official deception 110
oil drums 27
Old Testament 63, 64, 80
One Step Beyond (TV show) 220
One, the 53, 60
ONI 245
Open Skies, Open Minds (Pope) 194, 199
Operation Animal Mutilation 245
Operation Majestic-12 101
optical illusion 245
optical illusions 22, 139, 245
OPUS 211
orange 29, 30, 72, 75, 93, 117, 149, 187, 223
orange balls 72, 75
Orbiter 3 148
organic metal 236

Organization for Paranormal Understanding and Support . . 211
Organization for SETV Research 211
organizations . . 78, 91, 106, 110, 111, 113, 131, 189, 205, 343
original Blessed Lady of Ireland 2, 204
Orion 36, 62, 202
OSF 245
OSI 245
Osiris 62
OSS 245
ouranian 228, 246
out-of-body-experience 245
outer space . 29, 33, 34, 36, 44, 59, 65, 74, 80, 111, 114, 145,
154, 159, 164-166, 196, 198, 202, 213, 247
ovals 27, 129
oval-shaped craft 150
overflight 246
overpopulation 54, 86
oz 199, 213
O'Hare Airport 77
O'Neill, John 148
PAAPSI 211
Pacific Islands 21
Pacific Ocean 34, 166, 184, 238
painful medical tests 57, 93, 94
Pakistan 61
paleoanthropologists 54, 246
Paleolithic 61
pans 27
parachute canopy 246
parallel universes 37, 44, 169, 199
paranoia 90, 96
paranormal . 79, 206, 211, 213, 236, 245, 246, 249, 253, 343,
345

Paranormal and Alien Abduction Problem Solvers International 211
para-government 246
parhelia 240
parishioners 68
Parton, Dolly 344
party balloons 22
passenger jets 22, 184
passengers 24-26, 34, 142
pearl 29
Peking, China 25, 149
pellets 140
pencils 27
pendulum 33
Pentagon 73, 156, 197
personal mythology 177
pets 140
Pettus, Edmund W. 344
Phaeton 62
phantasmagoria 246
phenomenon . . . 10, 11, 28, 42, 61, 77, 79, 97, 101, 103, 107, 109, 113, 123, 129, 133, 137, 141, 146, 162, 168, 175, 188, 195, 197, 211, 216, 218, 227, 235, 239, 246, 252
Phenomenon Research Association 211
Phoenix UFO Incident 76
Phoenix, Arizona 75
PHOTINT 246
photocopies 190
photograph 188, 189, 191
photographed 40, 73, 129, 145, 148, 150, 162, 164, 165
photographic intelligence 246
photos . 4, 76, 78, 80, 116, 117, 122, 134, 136, 147, 148, 158, 160, 162, 164, 191, 221, 224, 241
physical evidence 140, 144, 189

physical scientists 177
physical sensations 58
physical world 169, 186
physics . . 28, 33, 35, 37, 44, 60, 71, 109, 148, 154, 155, 168, 169, 171, 175, 177, 214
pies 27
pillars of clouds 64
pillars of fires 64
Pillow, Gideon J. 344
pilot . . 28, 34, 48, 50, 72, 73, 80, 83, 136, 137, 143, 156, 159, 246
pilots . 10, 11, 25, 26, 29, 32, 58, 59, 71, 76, 77, 87, 111, 121, 125, 142, 143, 145, 151, 160, 168, 169, 234, 242
pink 29
planes 236
planet . . 29, 36, 46, 50, 59, 60, 62, 63, 74, 77, 81, 86, 89, 91, 96, 109, 139, 145, 153, 156, 163, 166, 167, 178, 190, 193, 201, 222, 224, 226, 232, 237, 244, 246, 248, 253
planets 22, 50, 113, 169, 170, 190, 232, 236, 237
plant samples 81
plant stalks 130
planting false magazine stories 126
planting false newspaper stories 126
planting false UFO stories 111
plants 141, 162, 184, 191
plastic 55
plates 27, 134
Pleiades 36
Pluto 246
Plutonian 246
Poland 209
Polaris missile 185
police . . . 11, 76, 102, 116, 120, 131, 136, 145, 168, 182, 189, 191, 247, 250

police officers 11, 76, 102, 120, 136, 145, 168
police reports 102
politicians 91
Polk, James K. 344
Polk, Leonidas 344
Polk, Lucius E. 344
pollution 54, 86, 226
poltergeist 202
polyhedrons 27
pop culture 59
Pope, Nick 9, 15, 152, 194, 197, 199, 342
populated areas 88, 102, 149, 167
populated cities 24
porthole 43
portholes 31, 104
Portugal 69, 209
positive spiritual feelings 96
power to heal 90, 132
PRA 211
praying mantis 48
Predator (movie) 59, 79
predators 97-99, 107
prehistoric art 143
prehistoric humanoid 47
prehistoric humans 246
prehistoric people 52, 62, 70, 79, 82
prehistoric primates 60
prehistoric times 61, 69, 132, 226
Presley, Elvis 344
priests 51
primates, nonhuman 54
primitive fish 174
prison 136, 143, 168, 237
prison sentences 143

private citizens 113, 124
probe 18, 87, 175, 176
probes 34, 164, 237, 248
professional investigators 99
Project U.F.O. (TV show) 220
Project Zare 207
proof . . . 84, 109, 139-145, 147, 152, 164, 169, 182, 212, 235
propulsion 35
propulsion system 33, 36
propulsion systems 143
protoplasm 247
prototype 247
psychic 90, 96, 202, 247
psychic abilities 90, 96
psychokinesis 251
psychology 167
Ptolemy 169
public sector 133
publicity 51
Puerto Rico 209
Puerto Rico UFO Network 209
Pulkovo Observatory 147
pulse rate 90
purple 29
pyramids 27, 53, 193, 202, 214, 241
Quay County, New Mexico 103
Quebec UFO Research 207
Quetzalcoatl 63
RAAF 28, 115, 247
rabbits 98, 140
radar . 11, 25, 29, 31, 33, 40, 73, 76, 105, 112, 122, 125, 139, 142, 145, 162, 169, 228, 245, 247, 248
Radar Approach Communications 247
Radar Approach Control 247

radar equipment 122
radar intelligence 247
radar records 125
radar screen 228
radar visual 248
radars 78, 142, 163, 169
radiation 11, 38, 39, 89, 106, 140, 247
radiation poisoning 106
RADINT 247
radio . . . 11, 22, 106, 111, 116, 125, 133, 170, 190, 191, 217, 218, 220, 247, 250, 344
radio detecting and ranging 247
radio station 116, 190
radio waves 22, 247
radioastronomy 111, 247
radios 37, 141, 233
RAF 247
RAF-USAF 38, 42, 185
rain 29, 43, 71, 164
rainbow 29
ranchers 97, 106, 114, 115
Randles, Jenny 196, 197, 200
Randolph, George W. 344
RAPCOM 247
RAPCON 247
Ravensdale, Cassidy 15
RCMP 247
Reagan, Ronald 152, 344
real science 171, 175, 177
record-keeping 247
red giant 50
refracted light 22
refrigerator 25
regression, hypnotic 88

religion 24
religious rituals 99
religious teachers 91
remotely piloted vehicle 31
Rendlesham Forest, Suffolk, England 39, 185
Reptilian 186
Reptilians 50
research stations 243
Reticulum 93
Revelation, book of 67
reverse-engineering 122, 156
Reynolds, Burt 344
riddles of UFOs 176
ridiculing UFO witnesses 111, 126
Riedel, Walter 155
riots 86
ripcord 246
Robbins, Hargus 344
Robbins, Peter 40, 199
Robert the Bruce, King 343
robot 46
robotic insects 85
robots 134
rocket 32, 67, 121, 154, 155, 159
rocket scientist 154
rockets 27, 142, 195, 235
rocket-like aircraft 67
Rocky Mountain UFO Conference 219
Roddenberry, Gene 79
Rogo, D. Scott 198
Romania 209
Romanian UFO Network 209
Romans 62
Rome 62, 149

Rome, Italy 150
Rosetta Stone 248
Roswell (TV show) 79
Roswell Army Air Field 115, 247
Roswell coverup 117
Roswell desert 115
Roswell Incident 114, 115, 117, 185, 196, 197, 226
Roswell, New Mexico 49, 114, 126, 185, 191
Roswellian 49
Roswellians 49
rotting vegetation 22
Roundtown UFO Society 211
Route 3 92
Royal Air Force 247
Royal Australian Air Force 247
Royal Canadian Mounted Police 131, 247
RPV 248
RPVs 31, 248
RQ-2 Pioneer 31
RQ-4 Global Hawk 31
Rucker, Edmund W. 344
runway 122
Russia . 21, 26, 55, 68, 146, 147, 149, 159, 163, 177, 230, 237
Russian 55
Russian church artists 68
Russian spectrograms 147
RV 25, 154, 248
RVs 25, 43
Ryder, Erin 15, 342
SA 248
SAC 248
SAC/HO 248
sacred energy paths 238
sacred sites 238

sacred UFO hot spots 185, 186
SAFOI 248
sailors 173
Saint Anthony of Alexandria 68
Saint Elmo's Fire 22, 125, 166
Saint Paul 66
San Francisco Expo West 219
Sand Dunes State Forest, Minnesota 185
Sanderson, Ivan T. 201
Santa Ana, California 136
Santa Claus, NASA code name for UFOs 6, 157
Santa Maria (ship) 28
SAO 248
Sao Paulo, Brazil 119
Sasquatch 202
SAT 248
satellite 92, 122, 148, 159, 166, 222, 224, 241, 248
satellite photos 122, 148
satellites 11, 22, 31, 141, 251
Saturn 22, 215, 248
Saturnian 248
saucer . . 27, 31, 43, 61, 64, 68, 69, 77, 82, 92, 115, 116, 124, 126, 130, 136, 140, 152, 157, 196, 200, 217, 231, 233, 235, 238, 248, 251
saucer nests 130
saucer-shaped UFOs 31, 238
saucerian 248
saucerinism 248
saucerite 249
saucerological 249
saucerology 249
saucers skipping over water 72
saving our planet 91
Scandinavia 52, 63, 70, 134, 235

Scandinavians 63, 70
scars 82, 96, 144
Schirra, Walter 6, 157
Schopenhauer, Arthur 178
science . 7, 9, 23, 46, 57, 65, 79, 107, 110-112, 148, 152, 156, 163, 166, 169-178, 181, 188, 195-197, 203, 205, 227, 230, 245
science book 112
Science Was Wrong (Friedman and Marden) 205
scientific knowledge 78, 112, 166
scientific method 171, 172
scientific pride factor 112
scientist . 45, 52, 113, 125, 152-155, 164, 170-172, 175, 177, 188, 193, 197
scientists . . 37, 50, 51, 61, 65, 68, 74, 75, 112, 144, 147, 148, 152, 156, 158, 159, 163, 164, 166-179, 184, 190, 198, 218, 227, 246, 248, 251, 343
scoop-mark 88, 144
scoop-marks 88, 96, 144
scorched 140
Scotland 69, 183, 191, 209, 343
Scottish 54
scramble 249
scratches 88, 144
screen memory 91, 249
Scruggs, Earl 344
scythes 27
sea mysteries, books on 201
sea serpents 202
Seabrook, John L. 344
Seabrook, Lochlainn 194, 201, 203, 204, 342-345
seaports 184, 191
Search for ExtraTerrestrial Intelligence 111, 249
searchlights 22, 43, 69, 71, 106, 142, 151

secrecy 116, 122, 154, 167, 228, 244
secret . . 10, 22, 45, 49, 71, 73, 81, 96, 99, 104, 106, 107, 110, 112, 114, 115, 117-123, 126, 127, 136-138, 140, 142-144, 154, 155, 163, 164, 166, 167, 169, 186, 193, 196, 198, 199, 201, 203, 204, 213-215, 227-230, 237-240, 245, 248
secret coded communication 229
secret conspiracy of silence 104
secret government 123, 126, 138, 144, 163, 199, 228
secret government agents 138
Secret Intelligence Service 239, 249
secret Nevada military base 45
secret Pentagon meetings 73
Secretary of the Air Force Office of Information 248
secretive 53
Security Alert Team 248
Security Police Squadron 250
Sedona, Arizona 186
Seger, Bob 344
self-correcting process, science as 175
self-propelled aircraft 70
self-luminescent 17, 30, 47
senior citizens 51
SETI 111, 112, 249
shadow 110, 211, 215, 249
shadow government 215, 249
shadow governments 110
shadows 145, 162, 241
Shag Harbour UFO Incident 29
Shag Harbour, Nova Scotia 29
shape . . . 26, 27, 30, 41-43, 59, 101, 104, 112, 121, 189, 221, 223, 233
shape-shift 26, 27, 43, 101, 104
Shard, the, on the Moon 148

shiny disks 73, 80
ship 28, 94, 168, 173, 242
ships 24, 25, 29, 58, 62, 68, 195, 242
shroud of secrecy 116
sick 140
Sightings (TV show) 220
SIGINT 249
Signals Intelligence 249
Signs (movie) 59
silent . . 22, 32, 43, 75, 99, 101, 102, 116, 122, 166, 200, 223
silent flying machines 99
silent UFOs 32
silver 29
silver disk 68
silver saucers 74
singularity 172
sinister quality 47
sinister-looking UFOs 158
Sinus Medii 148
Sirius 36, 62, 202, 209
Sirius UFO Space Sciences Research Center 209
SIS 239, 249
Skaggs, Ricky 344
Skandinavisk UFO Information 207
skeptic . . 7, 9-11, 17, 18, 23, 46, 50, 51, 57, 58, 74, 76-79, 85, 89, 92, 98, 100, 105, 118, 130, 132, 140, 143, 148, 149, 151, 152, 156, 163-168, 171, 174-179, 181, 190, 194, 196, 216, 236, 239, 245, 249
skeptic, defined 249
skeptical scientists 74, 178, 179, 190
skeptics . . 7, 9, 10, 46, 51, 74, 76-78, 85, 89, 92, 98, 100, 105, 118, 130, 132, 140, 143, 148, 149, 151, 152, 156, 163-168, 174-178, 194, 236, 239, 245
sketch pad 182

skin . . . 47, 55, 60, 66, 87, 89-91, 93, 99, 106, 119, 134, 135, 144
skin disorders 91
skin samples 87
sky 17, 18, 21-24, 29, 33, 35, 38-42, 61-63, 65-70, 72, 74, 75, 77, 80, 83, 92, 95, 106, 114, 118, 121, 122, 131, 135, 139, 143, 149-151, 165, 169, 170, 174, 181, 182, 186, 197, 198, 200, 201
sky galleons 68
sky ships 62
Skywatch Groningen 208
Skywatcher 249
slaughterhouse 100
Slayton, Donald 157
smell 55, 60, 85, 134
Smith, Wilbert B. 125, 133
Smithsonian Astrophysical Observatory 248
smoke 35, 86
smoking furnaces 64
snow 24, 71, 82, 97
snow-covered forests 24
Snowflake, Arizona 94
soap bubbles 27
Soares, Gerson de Macedo 155
Sociedade Portuguesa de Ovnilogia 209
Soft Support Building 250
soil 61, 81, 140
soil samples 81
solar system 44, 45, 50, 154, 170, 214, 241
solar systems 170, 190, 234
soldier 41, 42
soldiers 38-40, 51, 58, 68, 105, 119, 343
sonar 29, 245, 249, 250
sores 88

SOS OVNI Belgique 206
Souers, Sidney 117, 240
sound navigation and ranging 249
South America 98
South Ashburnham, Massachusetts 93
South Carolina 69, 344
South Dakota 97
South Korea 75, 229
South Pole 21, 186
Southend UFO Group 210
space 6, 7, 9, 11, 22, 27, 29, 33, 34, 36-38, 43-45, 47, 53, 59, 62-68, 70, 72, 74, 80, 82, 85, 87, 88, 94-96, 101, 102, 106, 110, 111, 114, 115, 121-123, 125, 139, 141, 143, 145, 146, 148, 151, 153, 154, 156-166, 168, 172, 184, 185, 191, 195-198, 202, 209, 213, 214, 227, 232, 237-239, 242, 243, 247, 250
Space Detection and Tracking System 250
space flight launching facilities 185, 191
space garbage 125
space matter 239
space shuttle 33, 34, 139, 161
space shuttle Discovery 139
Space Sister/Brother 250
space vehicles 6, 70, 227
space weapons 66
space-creature 68
spacecraft . 9, 11, 27, 34, 38, 45, 47, 63-65, 67, 74, 85, 87, 94-96, 101, 102, 106, 110, 114, 115, 121, 122, 145, 151, 156-158, 163, 165, 166, 184, 197, 242
SPADATS 250
Spain 149, 209
Spain's Air Ministry 124
sparks 35, 40
special agent 248

special agent in charge 248
Species (movie) 59
spectrograms 147
speed . . . 28, 33, 34, 37, 43, 44, 76, 134, 142, 145, 165, 166, 189, 221, 223, 225, 249-251
speed of light 33, 37, 44
sphere 17, 79, 82, 159, 227
Sphere (movie) 79
spheres 25, 27, 132, 140
spindles 27
spirals 27, 129
spirit 54, 122, 214
spirits 239
spiritual energies 131
spiritualism 250
spirituality 2, 203, 343
spirituality, and the aliens 86
Spock 163
spontaneous combustion 202
spotlights 31
spread of disinformation 126
spreading false rumors 111
Spring Heeled Jack 202
SPS 250
squadrons 73, 101, 103, 149, 150
squares 27, 129
squeaky voices 54
squid 172, 173
SR-71 Blackbird 122
SSB 250
star . 2, 23, 26, 37, 40, 48, 59, 92, 93, 122, 146, 163, 204, 222, 224, 237
Star Trek (TV show) 37, 48, 163, 237
Star Wars (movie) 37

Stargate (TV show) 79
stars . . . 22, 27, 36, 41, 92, 169, 170, 221, 224, 227, 234, 236
static 141
stealth 30, 122
stealthy UFOs 167
steam 35, 166
steam engine 166
Steiger, Brad 199, 203
Stephens, Alexander H. 344
Stewart, Alexander P. 344
stick-like bodies 48
stiffness 91
Stonehenge 186
stools 27
straight lines 99, 129, 241
Strand, Howard 154
strange craft 77, 92, 103, 157
strange lights 41, 80, 107
Strange magazine 217
Strategic Air Command History Office 248
street lights 22
Street, Dot 197
streetlights 22
Strieber, Anne 56
Strieber, Whitley 56-58, 199
Stringfield, Leonard 200
STS-51-A 159
Stuart, Jeb 344
subculture 250
submarines 31
submarine-shaped UFO 118
suborbit 250
SUFOI 207
SUFOR 210

Sumeria 65
Sumerian 65
Sun . . 22, 25, 50, 61, 62, 67-69, 81, 162, 169, 170, 221, 224, 239-241, 250
sun dog 250
sun dogs 240
sunglasses 46, 134
sunlight 17, 145, 232, 240
Sunna 52
super UFO 43
super UFOs 25
superior technology of aliens 114
supernatural 62, 137, 185, 203, 246
supernatural beings 137
supersonic 250
Supreme Being 52, 62
surgeon 98, 100
surgery 48, 57
surgical 98-100
surgical marks 98
surveying 251
suspended animation 94
swamp gas 22, 166
Sweden 69, 209
Swedenborg. Emanuel 69
Swindon UFO Research 210
swirled grass 140
Swiss astronomers 69
Switzerland 69, 151, 154, 209
swollen eyes 106
Syfy Channel 79
symbols 13, 103, 107, 118, 215
synagogue 185
S-4 45

tagging animals 90
tanks 58, 183
taped conversation 84
tapes 40, 111, 117, 125, 243
Tasmania 83
Tasmanian UFO Investigation Centre 206
Taylor, Richard 344
Taylor, Sarah K. 344
Taylor, Zachary 344
teardrops 27
Technical Report R-277 146
technological advantage 114
technology factor 114
telemeter 250
telemetry 250, 251
telepath 251
telepathic 55, 56, 60, 86, 94, 251
telepathic contactee 251
telepathically 55, 56, 60, 86, 94, 251
telepathy 54, 251
telepathy, used by aliens 86
telephone booth 25
telephone pole 23
telephones 141, 174
teleportation 67, 214, 251
telescope 80, 147, 182, 184, 187, 190
telescopes 78, 146
temperature inversion analysis 251
temperature inversions 22
temple 185, 202
temporary paralysis 91
Tennessee MUFON 211
tennis ball 25
Terminator (movie) 79

terrestrial 233
terrestrial vehicles 104
test planes 122
tetrahedral geometry 148
tetrahedrons 27
Texas 69, 80, 97, 105, 196
The Abyss (movie) 79
The Annunciation with Saint Emidius (Crivelli) 66
The Baptism of Christ (de Gelder) 175
The Bible and the Law of Attraction (Seabrook) 204
The Book of Kelle (Seabrook) 204
The Day After Roswell (Corso) 156, 197
The Day the Earth Stood Still (movie) 59
The Flying Saucers Are Real (Keyhoe) 164
The Forgotten (movie) 79
The Fourth Kind (movie) 79
The GAO Report on Roswell 197
The Goddess Dictionary of Words and Phrases (Seabrook) . 204
The Greek Myths (Graves) 204
The International UFO Library Magazine 217
The Invasion (movie) 59
The Madonna with Saint Giovannio (Ghirlandaio) 68
The National UFO Reporting Center 189
The Secret Jesus (Seabrook) 204
the supernatural 246
the truth . 10, 32, 51, 55, 82, 89, 110-113, 116, 117, 120-123, 126, 139, 143, 154, 175, 179, 197, 200, 215, 231, 245
The UFO Enigma 217
The UFO Experience Annual Conference 219
The UFOlogist 217
The Unexplained Files (TV show) 79
The Uninvited (Pope) 197
The War of the Worlds (movie) 59
The White Goddess (Graves) 205

The X-Files (TV show) 79
theodolite 251
Theory of Intelligent Force 130, 251
Thomson, William 170
thought disk 251
threads 140
Three Wise Men 66
thunderbirds 203
Thutmose III 61
TIA 251
TIF 130, 251
time . . 6, 17, 22-28, 30, 31, 33, 35-40, 44, 47, 50, 53, 57, 58, 68, 69, 71-75, 77, 80, 81, 85-88, 92, 96, 100, 101, 106, 109, 112, 116, 117, 123-125, 129, 135-137, 141, 142, 145-148, 150, 152, 154, 158, 162, 163, 168, 169, 171-174, 179, 181, 185, 189-191, 194, 195, 198, 199, 202, 214, 216, 218, 220-224, 227, 235, 236, 240, 242, 246, 344
time and space 37, 172, 202
tinsel 140
tinted windows 107
tired 94, 140
TLP 146, 147, 251
Tombaugh, Clyde 178
tongues 97
top secret . . 110, 115, 120, 121, 123, 126, 127, 193, 196, 199, 240
top secret document 123
top secret documents 240
top secret U.S. military base 121
Top Secret/Majic (Friedman) 199
tops 27, 95
torpedoes 27
torture 170

tourists 33, 78, 187
Tower, the, on the Moon 148
tractor 130, 131
traffic jams 149
translucent 30
transponder 90
trapezoids 27
TREAT 212
Triangle, the, on the Moon 148
triangles 27, 74, 129, 148, 193
triangular aircraft 27
triangular mark 87, 88
triangular-shaped UFO 149
Trouvelot, Etienne Leopold 146
truck 25
true science 171, 172, 177, 178, 181
Truman, Harry 21, 117, 152, 240
trumpets 27
truth . . 10, 16, 32, 51, 55, 81, 82, 89, 110-113, 116, 117, 120-123, 126, 139, 143, 154, 167-169, 171, 175, 178, 179, 197, 200, 210, 212, 214, 215, 231, 245
truth about aliens 111
Truth Seekers Midlands 210
TSA 77, 251
tsunamis 86
tubes 27, 241
TUFOIC 206
tunnel 94
Turkey 209
turquoise 29
Tuskegee experiments 117
TV . 11, 16, 40, 46, 49, 77-79, 86, 88, 93, 136, 141, 150, 165, 182, 189-191, 220, 343, 344
TV station 189-191

TVs 37, 141, 170, 174, 233
Twining, Nathan 117, 129, 240
Tynes, Ellen B. 344
U.S. . 4, 17, 21, 24, 32, 40, 42, 49, 59, 72, 73, 80, 81, 97, 99, 105, 109, 110, 113, 114, 116, 117, 119-123, 126, 127, 137, 138, 140, 141, 144, 152, 153, 155, 156, 164, 185, 187, 196, 197, 203, 228, 231, 234, 243, 250, 251, 343
U.S. Air Command 123
U.S. capital 72
U.S. Department of Defense 110
U.S. fighter jets 105
U.S. government . . 42, 59, 105, 109, 113, 114, 116, 117, 119, 123, 126, 127, 137, 138, 140, 144, 228, 231, 234
U.S. governmental denial 117
U.S. military 49, 73, 121, 122, 187
U.S. military authorities 121
U.S. Navy 59, 123, 155
UAO 251, 252
UAP 252
UAV 252
UAVs 31
UFO . 7, 9-11, 21-59, 62-95, 97-107, 109-126, 129-131, 133-137, 140-159, 162-179, 181-191, 193-201, 205-224, 226, 228-233, 235-240, 242-245, 248, 249, 252, 253
UFO & Paranormal Research Society of Australia 206
UFO activity 104, 141
UFO amateur radio stations 217
UFO books 194
UFO clubs 205
UFO conferences 78, 216, 218, 220
UFO conferences, list of 218
UFO Contact Center International 212
UFO coverup . 7, 109, 110, 114, 122, 124, 126, 137, 144, 199
UFO Coverup, books on the 199

UFO crashes, books on 197
UFO document 213
UFO documents 78, 229
UFO Encounters magazine 217
UFO enthusiasts 236
UFO era 37, 70, 72, 187
UFO Experience Support Association 206
UFO eyewitnesses 152, 232
UFO files 124, 149, 199, 213, 220
UFO Files (TV show) 220
UFO groups 205
UFO Hunters (TV show) 220
UFO Magazine 217
UFO magazines 78, 216, 220
UFO magazines, list of 216
UFO Newsclipping Service 216
UFO organization 118, 136, 189
UFO organizations 78, 111, 205
UFO organizations, infiltration of 111, 126
UFO phenomenon 235
UFO reality 196, 249
UFO Report 109, 188, 217
UFO Reporter 217
UFO reports 25, 28, 38, 78, 80, 183
UFO Research Association of Ireland 208
UFO Research Queensland 206
UFO Resource Center 211
UFO sighting . 7, 21, 40, 76, 137, 182, 189-191, 195, 221, 223, 238, 242
UFO sighting report form 7, 190, 191, 223
UFO sightings 237
UFO Society of Ireland 208
UFO technology 114, 117, 122, 240
UFO watcher 7, 181-183, 188, 190, 191, 249

UFO watchers 182, 183, 186, 191, 194, 235
UFO watching group 183
UFO watching site 183
UFO Werkgroep Nederland 208
UFO Wisconsin 211
UFO witnesses . 9, 24, 65, 111, 126, 134, 136, 140, 164, 167, 168
UFO-FBI Connection (Maccabee) 199
UFO-Finland 207
UFO-related Websites 216
UFO-Rogaland 208
UFO-Sweden 209
ufocal 252
UFOCCI 212
UFOESA 206
UFOIC 206
ufoism 252
UFOlats 208
ufological 194, 252
ufologist 7, 38, 94, 193, 194, 205, 217, 219, 220, 252
ufologists . . . 10, 34, 38, 44, 46, 81, 103, 119, 120, 122, 124, 144, 163, 185, 186, 194, 215, 216, 218-220, 234
ufology . . 9, 10, 56, 78-80, 150, 188, 194, 219, 220, 225, 230, 231, 244, 248, 252, 253, 343
ufonauts 252
UFORCE 211
UFOria 252
UFOs . 1-7, 10, 11, 16-19, 21-38, 42-48, 50-52, 56-58, 61-64, 66-70, 72-81, 84, 86-88, 93, 98, 100-105, 107, 109-114, 116-118, 120-127, 130-133, 135-160, 162-172, 174-179, 181-189, 191, 193-203, 206, 207, 209, 210, 212, 213, 216, 218-221, 223, 226-228, 230-240, 242-245, 248-252

UFOs and aliens . 2, 4, 10, 16-18, 44, 56, 58, 61, 63, 68, 78-80, 100, 103, 110, 112-114, 116, 118, 123, 125, 126, 144, 147, 150, 164, 167, 176, 181, 183, 193, 194, 216, 220, 221, 231, 234, 235, 249, 252
UFOs and Aliens: The Complete Guidebook (Seabrook) . . . 194
UFOs Over Earth (TV show) 220
UFOSA 206
UFOzarks 211
UFO-detecting equipment 80
UFO-related documents 123, 125
UGM 252
UGMs 129
UK 209-211, 217, 342
Ukert crater 148
Ukert region 148
ultradimensional 252
ultraviolet light, UFOs operating in 35
undefined sensory experience 252, 253
underground 36, 110, 117, 119, 122, 186, 200, 215
underground military installations 110
underwater 27-29, 31, 36, 43, 44, 181, 249, 252, 253
underwater unidentified object 252
unexplained aerial object 252
unexplained mysteries, books on 201, 203
Unhuman Flying Objects South Australia 206
unidentifiable flying object 42
unidentified flying object 21, 74, 157, 181, 219, 242, 253
unidentified flying objects . 18, 23, 25, 29, 42, 70, 78, 88, 113, 120, 153, 171, 194, 205, 229
unidentified submerged object 27
unidentified underwater object 28
uninhabited regions 24
Union County, New Mexico 103
United Airlines 77

United Kingdom 11, 120, 144, 209
United Nations 156
United States Air Force 253
United States Government 253
Universe . . 36, 37, 53, 56, 111, 113, 114, 164, 166, 167, 169-174, 176, 177, 190, 202, 212, 232, 234
University of Campinas 119
Unknown Phenomena Investigation Association 210
unmarked helicopters 101, 102, 106, 107
unusual ground marking 252
Uranian 253
Uranus 253
USA . . . 3, 4, 18, 38, 40, 42, 51, 72, 109, 125, 154, 167, 185, 211, 212, 216-219, 223, 224, 253
USAF 38, 40, 42, 72, 109, 125, 154, 167, 185, 253
USE 253
USG 253
USSR 155
UUO 28, 252, 253
UUO case 28, 29
Uzbekistan 165
Valentich, Frederick 83, 84
Valkyries 63
Vallee, Jacques 198, 200
Vance, Robert B. 344
Vance, Zebulon 344
Vandenberg Air Force Base 184
Vandenberg, Hoyt 117, 240
vaporize 239
Varginha Coverup 120
Varginha, Brazil 118
Vatican 113, 149, 150
Vatican theologian 113
vegetables 46

vegetarian 60
vegetarians 55, 56, 90
Venable, Charles S. 344
Venus 22, 52, 63, 190, 253
veterinarians 98
Victoria and Albert Museum 154
Victorian UFO Research Society 206
videotape 116
videotaped 77
video-recorder 191
Vietnam 234
Vietnam War 71, 74, 105
Viking probe 237
violence 86
violent species 60
Virgin Mary 52, 67, 82, 83
Virginia 17, 52, 69, 165, 229, 343
visions 199, 205
Visoki Decani fresco 67
vitreous floater 23
vivid dreams 84
von Braun, Wernher 154
vortex 185-187, 200, 211, 253
vortex center 186, 253
vortex centers, books on 200
Vortex Society 211
Voskhod 1 159
Voyager 34, 164, 237
Voyager probe 237
Voyager probes 34, 164, 237
VS 211
VUFORS 206
V-shaped squadrons 150
V-shaped UFO 75

Wales 135, 217
Walker, Joseph A. 159
walls 56, 61, 62, 85
walnuts 27
Walton, Travis 94, 95, 198
war . 2, 54, 58, 59, 71, 72, 74, 80, 81, 86, 105, 117, 154, 155, 183, 201, 226, 234, 245, 343, 344
warehouses 122
Warren, Larry 40, 199
Washington D.C. 72
Washington merry-go-round 72
Washington Monument 75
Washington Post 73
Washington, John A. 344
Washington, Thornton A. 344
watches 88, 92
water 17, 26-28, 33, 72, 145, 162, 175, 190, 214, 230
Watergate 74
Watergate Break-in 117
weather . 9, 11, 22, 24, 71, 114-116, 121, 125, 130, 142, 157, 164, 166, 214, 221, 223, 225, 252
weather balloons 11, 22, 114, 125
Webb, Walter N. 178
weight loss 106
weird mist 92
weirdness 40
Welsh 54
West Virginia 52
Western Europe 98
Western states 97
Westinghouse 153
Weston Fireball 174
whale 60
whalers 173

whales 90
wheatfield 126
wheel 53, 65, 150
wheels 65, 66, 139
Wheless, Hewitt T. 137
whine 32
whirlwind 64, 65, 130
white . 17, 29, 30, 33, 43, 47, 58, 75, 76, 79, 80, 92, 134, 140, 159, 165, 205, 229
White House 58, 79, 80
White Mountains 92
White Sands Proving Grounds 33
white shirts 134
white skin 47
White, Robert 159
wild animals 90, 98, 140
Wilkens, H. P. 148
Wilkins, H. Percy 178
Wilkins, Harold T. 203
Will-O'-the-Wisp 22
Wilson, Woodrow 344
Wiltshire Times 129
Wiltshire, England 75, 129
Winder, Charles S. 344
Winder, John H. 344
window 17, 31, 43, 85, 93, 157, 165, 228
windows 23, 31, 32, 43, 56, 92, 103, 107, 142, 224
wing-shaped UFOs 27
wings . 4, 27, 36, 62, 63, 72, 80, 125, 142, 151, 224, 226, 231
Wisconsin 97, 211
Witherspoon, Reese 344
wobble 33
wolves 90, 98, 99, 107
Womack, John B. 344

Womack, Lee A. 344
woman 48, 61, 116, 119, 135, 205
women 49, 83, 135, 172, 178, 203
Women's Movement 74
wood 38, 40, 42, 44, 55, 94, 140, 151, 185, 199
woods 18, 39, 41, 95, 119, 140
world . 7, 11, 17, 21, 23, 32, 38, 48, 50, 51, 53, 54, 57-60, 67, 70-72, 75, 77, 78, 80, 81, 83, 84, 86, 94, 103, 106, 107, 109, 110, 113, 114, 116, 118, 120, 122, 124, 126, 129, 131, 133, 139, 143, 145, 150, 154-156, 162, 166, 169-171, 174, 176, 185, 186, 193, 195, 196, 198, 201, 203, 204, 214, 215, 220, 228, 234, 236, 238, 245, 249, 343
World War II 71, 72, 80, 154, 155, 234, 245
worlds 59, 72, 201, 203, 253
worldwide coverup of UFOs 126
Worldwide UFO Newsclipping Bureau 216
wormholes 44
worshiping demons 231
wrap-around sunglasses 46
wreckage 105, 115, 116
Wright brothers 69, 146
Wright Field 115
Wright, Orville 70
Wright, Wilbur 70
writing 17, 53, 70, 115, 118, 230, 343
X-rays 170
X-15 159
X-43A Scramjet 34
X-files 29, 79, 253
X-ray 89
Yahweh 237
Yakima, Washington 187
Yegorov, Boris 159
yellow 26, 29, 39, 41

yellow fog 39
Yeti 202
Yugoslavia 67
Zechariah 65
Zerotime UFO Research 211
Zeta 1 92
Zeta 2 92
Zeta Reticuli 92, 93
Zeus 62, 237
zig-zag 33
Zigel, Felix 177
Zimbabwe, South Africa 150
Zollicoffer, Felix K. 344
zoo 119
zoology 202
Zulu 218, 253

WEBSITES

FOR THE PEOPLE CONNECTED WITH THIS BOOK

- Lochlainn Seabrook: www.lochlainnseabrook.com
- Nick Pope: www.nickpope.net
- Stanton T. Friedman: www.stantonfriedman.com
- Timothy Good: www.timothygood.co.uk
- Dr. Bruce Maccabee: www.brumac.8k.com
- Erin Ryder: www.erinryder.com

MEET THE AUTHOR

LOCHLAINN SEABROOK, winner of the prestigious Jefferson Davis Historical Gold Medal for his "masterpiece," *A Rebel Born: A Defense of Nathan Bedford Forrest*, is an unreconstructed Southern historian, award-winning author, Civil War scholar, and traditional Southern Agrarian of Scottish, English, Irish, Welsh, German, and Italian extraction. An encyclopedist, lexicographer, musician, artist, graphic designer, genealogist, and photographer, as well as an award-winning poet, songwriter, and screenwriter, he has a 40 year background in historical nonfiction writing and is a member of the Sons of Confederate Veterans and the National Grange.

Due to similarities in their writing styles, ideas, and literary works, Seabrook is often referred to as the "new Shelby Foote," the "Southern Joseph Campbell," and the "American Robert Graves" (his English cousin).

The grandson of an Appalachian coal-mining family, Seabrook is a seventh-generation Kentuckian, co-chair of the Jent/Gent Family Committee (Kentucky), founder and director of the Blakeney Family Tree Project, and a board member of the Friends of Colonel Benjamin E. Caudill.

Seabrook's literary works have been endorsed by leading authorities, museum curators, award-winning historians, bestselling authors, celebrities, noted scientists, well respected educators, TV show hosts and producers, renowned military artists, esteemed Southern organizations, and distinguished academicians from around the world.

Seabrook has authored over 40 popular adult books on the American Civil War, American and international slavery, the U.S. Confederacy (1781), the Southern Confederacy (1861), religion, theology and thealogy, Jesus, the Bible, the Apocrypha, the Law of Attraction, alternative health, spirituality, ghost stories, the paranormal, ufology, social issues, and cross-cultural studies of the family and marriage. His Confederate biographies, pro-South studies, genealogical monographs, family histories, military encyclopedias, self-help guides, and etymological dictionaries have received wide acclaim.

Seabrook's eight children's books include a Southern guide to the Civil War, a biography of Nathan Bedford Forrest, a dictionary of religion and myth, a rewriting of the King Arthur legend (which reinstates the original pre-Christian motifs), two bedtime stories for preschoolers, a naturalist's guidebook to owls, a worldwide look at the family, and an examination of the Near-Death Experience.

Of blue-blooded Southern stock through his Kentucky, Tennessee, Virginia, West Virginia, and North Carolina ancestors, he is a direct descendant of European royalty via his 6th great-grandfather, the Earl of Oxford, after which London's famous Harley Street is named. Among his celebrated male Celtic ancestors is Robert the Bruce, King of Scotland, Seabrook's 22nd great-grandfather. The 21st great-grandson of Edward I "Longshanks" Plantagenet), King of England, Seabrook is a thirteenth-generation Southerner through his descent from the colonists of Jamestown, Virginia (1607).

The 2nd, 3rd, and 4th great-grandson of dozens of Confederate soldiers, one of his closest connections to the War for Southern Independence is through his 3rd great-grandfather, Elias Jent, Sr., who fought for the Confederacy in the Thirteenth Cavalry Kentucky under Seabrook's 2nd cousin, Colonel Benjamin E. Caudill. The Thirteenth, also known as "Caudill's Army," fought in numerous conflicts, including the Battles of Saltville, Gladsville, Mill Cliff, Poor Fork, Whitesburg, and Leatherwood.

Seabrook is a descendant of the families of Alexander H. Stephens, John Singleton Mosby, and Edmund Winchester Rucker, and is related to the following Confederates and other 19th-Century luminaries: Robert E. Lee, Stephen Dill Lee, Stonewall Jackson, Nathan Bedford Forrest, James Longstreet, John Hunt Morgan, Jeb Stuart, P. G. T. Beauregard (designed the Confederate Battle Flag), George W. Gordon, John Bell Hood, Alexander Peter Stewart, Arthur M. Manigault, Joseph Manigault, Charles Scott Venable, Thornton A. Washington, John A. Washington, Abraham Buford, Edmund W. Pettus, Theodrick "Tod" Carter, John B. Womack, John H. Winder, Gideon J. Pillow, States Rights Gist, Henry R. Jackson, John Lawton Seabrook, John C. Breckinridge, Leonidas Polk, Zachary Taylor, Sarah Knox Taylor (first wife of Jefferson Davis), Richard Taylor, Davy Crockett, Daniel Boone, Meriwether Lewis (of the Lewis and Clark Expedition) Andrew Jackson, James K. Polk, Abram Poindexter Maury (founder of Franklin, TN), William Giles Harding, Zebulon Vance, Thomas Jefferson, George Wythe Randolph (grandson of Jefferson), Felix K. Zollicoffer, Fitzhugh Lee, Nathaniel F. Cheairs, Jesse James, Frank James, Robert Brank Vance, Charles Sidney Winder, John W. McGavock, Caroline E. (Winder) McGavock, David Harding McGavock, Lysander McGavock, James Randal McGavock, Randal William McGavock, Francis McGavock, Emily McGavock, William Henry F. Lee, Lucius E. Polk, Minor Meriwether (husband of noted pro-South author Elizabeth Avery Meriwether), Ellen Bourne Tynes (wife of Forrest's chief of artillery, Captain John W. Morton), South Carolina Senators Preston Smith Brooks and Andrew Pickens Butler, and famed South Carolina diarist Mary Chesnut.

Seabrook's modern day cousins include: Patrick J. Buchanan (conservative author), Cindy Crawford (model), Shelby Lee Adams (Letcher County, Kentucky, portrait photographer), Bertram Thomas Combs (Kentucky's fiftieth governor), Edith Bolling (wife of President Woodrow Wilson), and actors Robert Duvall, Reese Witherspoon, Lee Marvin, Rebecca Gayheart, Andy Griffith, and Tom Cruise.

Seabrook's screenplay, *A Rebel Born*, based on his book of the same name, has been signed with acclaimed filmmaker Christopher Forbes (of Forbes Film). It is now in pre-production, and is set for release in 2016 as a full-length feature film. This will be the first movie ever made of Nathan Bedford Forrest's life story, and as a historically accurate project written from the Southern perspective, is destined to be one of the most talked about Civil War films of all time.

Born with music in his blood, Seabrook is an award-winning, multi-genre, BMI-Nashville songwriter and lyricist who has composed some 3,000 songs (250 albums), and whose original music has been heard in film (*A Rebel Born, Union Bound, Cowgirls 'n Angels*) and on TV and radio worldwide. A musician, producer, multi-instrumentalist, and renown performer—whose keyboard work has been variously compared to pianists from Hargus Robbins and Vince Guaraldi to Elton John and Leonard Bernstein—Seabrook has opened for groups such as the Earl Scruggs Review, Ted Nugent, and Bob Seger, and has performed privately for such public figures as President Ronald Reagan, Burt Reynolds, Loni Anderson, and Senator Edward W. Brooke. Seabrook's cousins in the music business include: Johnny Cash, Elvis Presley, Billy Ray and Miley Cyrus, Patty Loveless, Tim McGraw, Lee Ann Womack, Dolly Parton, Pat Boone, Naomi, Wynonna, and Ashley Judd, Ricky Skaggs, the Sunshine Sisters, Martha Carson, and Chet Atkins.

Seabrook lives with his wife and family in historic Middle Tennessee, the heart of Forrest country and the Confederacy, where his conservative Southern ancestors fought valiantly against Liberal Lincoln and the progressive North in defense of Jeffersonianism, constitutional government, and personal liberty.

LOCHLAINNSEABROOK.COM

If you enjoyed this book you will be interested in Mr. Seabrook's works on the paranormal, myth, and religion:

- ☛ CARNTON PLANTATION GHOST STORIES
- ☛ CHRISTMAS BEFORE CHRISTIANITY
- ☛ JESUS AND THE LAW OF ATTRACTION
- ☛ BRITANNIA RULES

Available from Sea Raven Press and wherever fine books are sold

ALL OF OUR BOOK COVERS ARE AVAILABLE AS 11" X 17" POSTERS, SUITABLE FOR FRAMING

SeaRavenPress.com

CPSIA information can be obtained
at www.ICGtesting.com
Printed in the USA
BVOW11s1302230917
495663BV00001B/3/P